M

DVD Players and Driv

DVD Players and Drives

K.F. Ibrahim
Senior Lecturer
College of North West London

Newnes

AMSTERDAM BOSTON HEIDELBERG LONDON NEW YORK OXFORD
PARIS SAN DIEGO SAN FRANCISCO SINGAPORE SYDNEY TOKYO

Newnes
An imprint of Elsevier
Linacre House, Jordan Hill, Oxford OX2 8DP
200 Wheeler Road, Burlington, MA 01803

First published 2003

British Library Cataloguing in Publication Data
A catalogue record for this book is available from the British Library

Library of Congress Cataloging in Publication Data
A catalog record for this book is available from the Library of Congress

ISBN 0 7506 5736 7

For information on all Newnes publications
visit our website at www.newnespress.com

Typeset by Replika Press Pvt Ltd, India
Printed and bound in Great Britain by Biddles Ltd

Contents

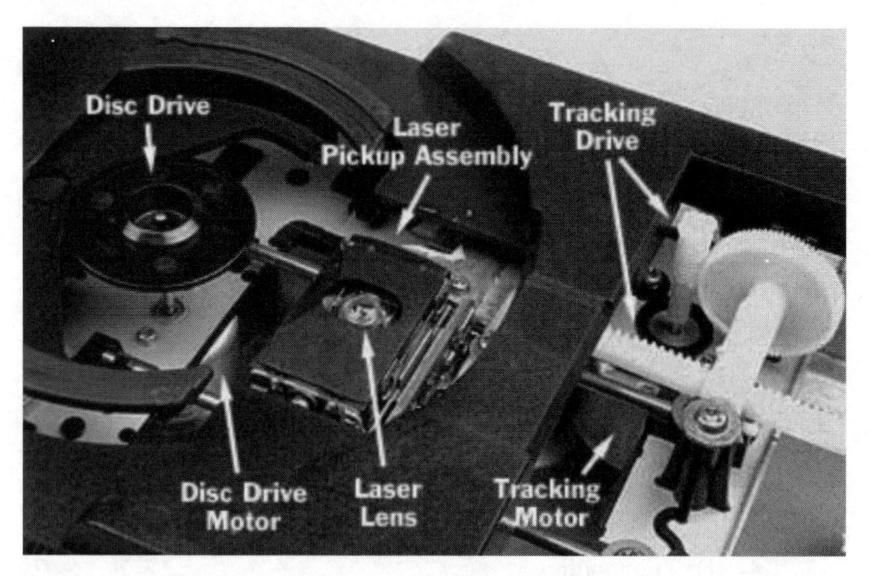

The picture shows the optical/electromagnetic assembly of a DVD player. It shows the laser pickup head together with its associated control mechanism: tracking, focus, and motor drives

PREFACE

Digital video technology has, in the past few years, experienced an unprecedented expansion, exceeded only by the mobile phone technology and today the two are merging. Digital television is now common-place, DVD players are replacing the old VHS players, and digital video recording is becoming more popular. Knowledge of digital video technology has become an important part of electronic engineering.

The Introduction in Chapter 1 provides an overview of DVD discs construction and properties and an overview of the encoding and decoding (i.e. playback) systems. The necessary background knowledge of digital and microprocessor applications including error correction techniques is provided in Chapter 2. Chapters 3 to 5 deal with the encoding process in detail and Chapters 6 to 11 cover the decoding process, including practical applications in terms of circuit diagrams of DVD players. Of particular interest to servicing engineers will be Chapter 10, which deals with faults and fault-finding techniques on DVD players.

An introduction to the production of DVD titles is contained in Chapter 12 and the installation and configuration of DVD drives is covered in Chapter 13. The latter commences with an overview of PC technology for those who are not familiar with personal computers.

The book contains a comprehensive glossary and five appendices including a set of self-test questions in Appendix E.

I have written the book with electronic students as well as practicing engineers in mind. It is suitable for the digital video part of craft and technician courses and I hope will be of some considerable assistance to servicing and development engineers.

K. F. Ibrahim
July 2003

Chapter 1

Introduction

DVD stands for digital versatile disc or, more commonly, digital video disc. DVDs are the most recent generation of compact disc technology, and are able to store huge amounts (gigabytes) of data – comparable to the hard disc commonly used in computers. DVDs are available in two diameter sizes; 12 cm and 8 cm. The 12 cm DVD is the same size as the familiar CD (audio or ROM), but holds over 25 times more data. This takes the DVD to a qualitatively higher level in terms of its applications. For the first time, we have a compact disc that can hold over 2 hours of high quality video with six channels of high-fidelity surround sound. As well as being a mass storage device for computer applications and educational/ training purposes that make use of interactive facilities, other applications include archiving of books and still pictures.

The first DVD format, the DVD-video was introduced in 1996 with its specifications published in *Book B*. 1 Previous to that, optical disc technology was limited to the compact disc (CD) and the Laserdisc. The CD had two main formats, audio CD and CD-ROM. The Laserdisc was 30 cm in diameter, and was intended to hold high quality audio and video of up to 1 hour per side. That technology never caught on and was quickly superseded by the far superior DVD.

The DVD-video was joined by the DVD-ROM format (*Book A*) and these were followed by other standards – including DVD-Audio, DVD-R, DVD-RAM and DVD-RW (see Table 1.1).

Table 1.1 DVD formats

DVD-ROM	Read only	*Book A*
DVD-video	Video	*Book B*
DVD-audio	Multi-channel	*Book C*
DVD-R	Recordable	*Book D*
DVD-RAM	Random access	*Book E*
DVD-RW, DVD+RW	Rewriteable	*Book F*

The DVD

The basic principle of storing data on a digital versatile disc (DVD) is the same as in the audio CD and the CD-ROM – namely the creation of *pits* and *lands*, which are translated into ones and zeros by a laser pickup head. By using a laser beam with a shorter wavelength than the conventional CD, the pits could be smaller and thus the storage capacity increased to few gigabytes (GBs). The shorter wavelength (650 or 635 nm, compared with 780 nm for the CD) reduces the size of the beam spot, thus allowing the laser to focus on a smaller pit size and a closer track pitch (Figure 1.1). Improved signal processing and error correction techniques also contribute to the increase in data capacity.

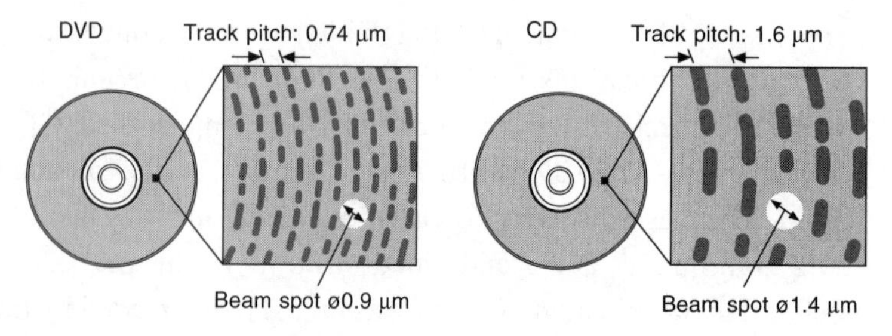

Figure 1.1

Unlike the CD, which has one substrate, the DVD consists of two 0.6-mm thick substrates bonded back-to-back which gives it the necessary stiffness to avoid disc wobble or tilt.

DVDs may record data on one side only (single-sided) or on both sides (double-sided). Each side may have a single layer of recording or two layers (dual-layer) as illustrated in Figure 1.2. This gives five different construction formats:

- Single-sided (SS), single-layer (SL)
- Single-sided (SS), dual-layer (DL)
- Double-sided (DS), single-layer (SL)
- Double-sided (DS), dual-layer (DL) on one side only
- Double-sided (DS), dual-layer (DL) on both sides.

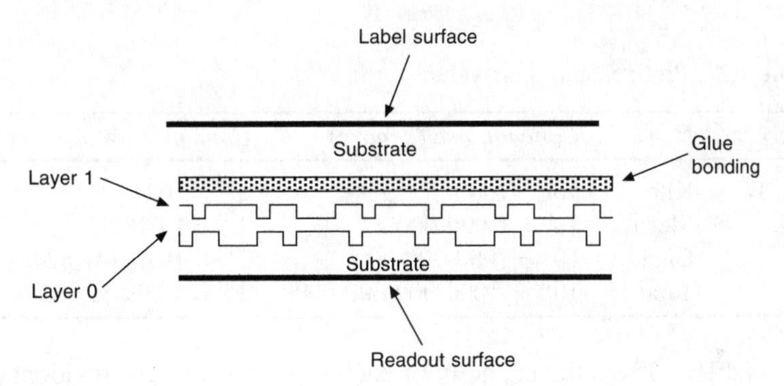

Figure 1.2
DVD Disc Construction

With dual layer construction, layer 1 must be read through layer 0. Hence layer 0 must be semi transparent, with a reflection of 20 per cent, while the second layer must have a reflection of 70 per cent. For double-sided discs, data must be read from both sides of the disc.

DVD capacity or size

The size or capacity of a DVD is the number of bits of data it can store – that is, the number of pits and lands it can accommodate.

Capacity is specified as so many bytes, in the same way as with a computer storage device such as a hard disk. When the capacity exceeds thousands, millions or billions of bytes (as it often does), the prefixes K (or k), M or G respectively are used. In ordinary usage, K = 1000 (10^3), M = 1000K (10^6) and G = 1000M (10^9). In specifying a computer data capacity a 'binary' kilo, mega and giga are used, which are based on powers of 2 – namely 2^{10} for kilo, 2^{20} for mega and 2^{30} for giga. This makes a computer K = 1024 instead of the commonly used denary 1000, and so on (see Table 1.2). In DVD applications, using computer-based data capacities is not relevant. Nonetheless, it is not uncommon for these to be used in publications and specifications. For the purposes of this book, the commonly used denary prefixes will be used unless otherwise stated.

Table 1.2 Prefixes and their values

Prefix	*Name*	*Common use (denary)*	*Computer use (binary)*
K or k	Kilo	10^3 = 1000	2^{10} = 1024
M	Mega	10^6 = 1 000 000	2^{20} = 1 048 576
G	Giga	10^9 = 1000 000 000	2^{30} = 1073 741 824
T	Tera	10^{12} = 1000 000 000 000	2^{40} = 1 099 511 628

Table 1.3 lists the capacity of each type of DVD. It is evident that the capacity of a DVD does not double with dual disc configuration compared with the single-layer type. This is because a longer pit length is used for the dual-layer type to avoid reading errors caused by interference between the layers.

Disc layout

The DVD disc is divided into four areas, as shown in Figure 1.3. The main area is the user or data area, where video and audio data are recorded. This is preceded by a *lead-in area* including control data areas, and is followed by a *lead-out area*. Before the video and audio data are recorded in the user area, the data bits are

Table 1.3 DVD disc storage capacity

12-cm discs

Name	*Type*	*Capacity in billions of bytes*	*Capacity in binary gigabytes*	*Approximate recording time in hours of video*
DVD-5	SS/SL	4.7	4.38	2.25
DVD-9	SS/DL	8.54	7.98	4
DVD-10	DS/SL	9.4	8.75	4.5
DVD-14	DS/DL on one side only	13.24	12.33	6
DVD-18	DS/DL on both sides	17.08	15.9	8

8-cm discs

Type	*Capacity in billions of bytes*	*Capacity in binary gigabytes*	*Approximate recording time in hours of video*
SS/SL	1.43	1.36	0.75
SS/DL	2.6	2.48	1.25
DS/SL	2.85	2.72	1.5
DS/DL	5.19	4.95	2.5

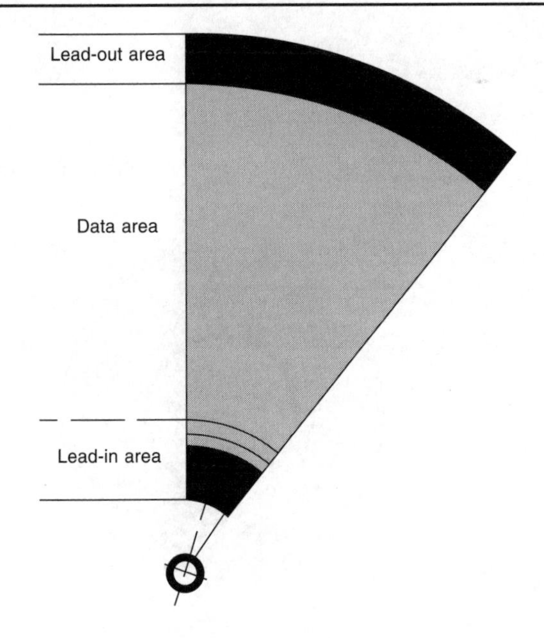

Figure 1.3

grouped into 2048-byte chunks known as *sectors*. Each sector commences with a header, which includes (among other things) a unique ID number for the sector. The data are written on the disc surface, sector by sector. Reading the data off the disc involves directing the pickup head to the part of the disc surface where the required sector is recorded, and once the sector has been identified, data are extracted for further processing.

Burst cutting area

The burst cutting area is a section near the hub of the disc (Figure 1.4). It is invariably used for stamping up to 188 bytes of information related to the individual disc, such as ID codes, serial number, or any other information that may be used for inventory purposes or storage systems such as jukeboxes for quick identification of the individual disc. DVD specification states that the burst cutting area must be between 44.6 and 47.0 mm from the centre of the disc. The burst cutting area may be read by the same pickup head that is used for reading the user data off the disc.

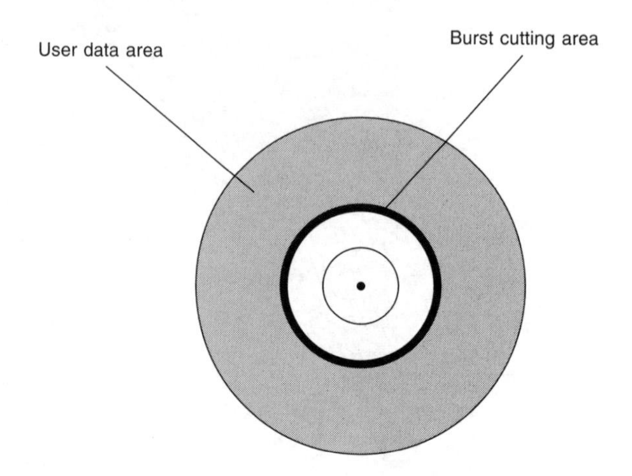

Figure 1.4

Recordable discs

The original DVD specification related to the read-only (non-recordable) *DVD-ROM*. Today, there are different recordable formats for DVD applications:

- Write-once formats – *DVD-R* and *DVD+R*, R for recordable
- Rewritable formats – *DVD-RAM*, *DVD-RW* and *DVD+RW*.

The various recording formats have different features. DVD-R and DVD+R are both compatible with over 80 per cent of all DVD players and DVD-ROM drives. DVD-RW was the first DVD recording format released that is compatible with almost 70 per cent of all DVD players and DVD-ROM drives; it supports single-sided 4.7-GB and double-sided 9.4-GB disc capacities. DVD+RW has better features than DVD-R, such as lossless linking, constant angular velocity (*CAV*) writing, and constant linear velocity (*CLV*) writing. Its greatest advantage is that it is compatible with almost all DVD players and DVD-ROM drives. It supports single-sided 4.7-GB and double-sided 9.4-GB disc capacities. DVD-RAM has the best recording features, but unfortunately it is incompatible with most DVD players and DVD-ROM drives. DVD-RAM discs are used more as removable storage devices than as a recording medium for audio/video information.

There are two versions of DVD-R format: DVD-R(G) for General and the DVD-R(A) for Authoring. It is also possible to create a hybrid disc that is partially read-only and partially recordable, sometimes known as DVD-PROM. This would normally be a disc with two layers, one of which is read-only while the other is a recordable part.

While a DVD-ROM disc such as a DVD-video disc has pits stamped permanently onto its surface, writable or recordable discs use other techniques to produce the effect of a pit (i.e. reducing the strength of the reflected laser beam). In the case of a DVD-R disc, a photosensitive dye is used to cover a reflective metallic surface. When the dye is heated by a high-power (6–12 mW) pulsating laser beam, it becomes darker and hence less transparent. A pulsating laser is necessary to avoid overheating the dye and thus creating oversized pits. The disc is manufactured with a *wobbled groove*, which is moulded into the substrate and provides a self-regulating clock to guide the laser beam as it burns the disc. The wobbled track is pre-divided into sectors, and each sector is identified by a pre-stamped header. DVD-R is available in two versions: version 1 has a capacity of 3.95 billion bytes (3.68 binary GB), and version 2 has a capacity of 4.7 billion bytes (4.37 binary GB).

DVD-RAM is an erasable and recordable version of DVD-ROM. While employing the same wobbled groove as DVD-R it uses a different technology, known as *phase change*. Phase-change technology uses a metal compound that alters its reflectivity as it changes between a crystallized and an amorphous state (Figure 1.5). When the compound is heated by a low-power laser it melts, creating a crystallized spot with a high reflectivity known as '*eraser*'; alternatively, if it is heated by a high-power laser it melts and then cools down rapidly to form an amorphous (non-crystalline) spot of low reflectivity, known as '*mark*'. The marks can be read by a low-power laser, to retrieve the data from the disc. Phase-change uses the same wobbled groove as DVD-ROM, but it writes its data bits

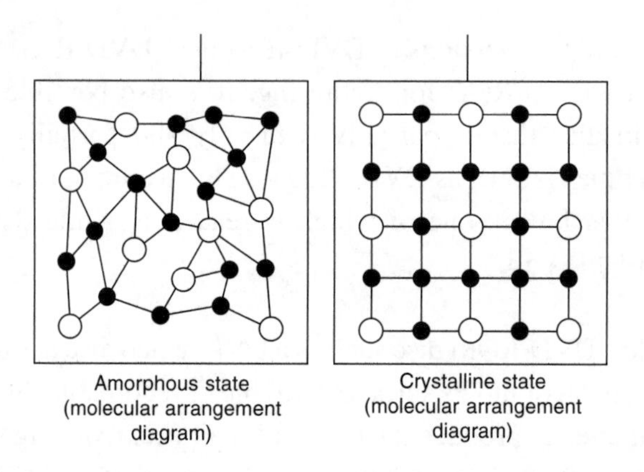

Figure 1.5

in both the groove and the land between the grooves (see Figure 1.6). The purpose of the wobbled groove is twofold:

1. It generates a spindle motor control signal
2. It generates a gate signal, which is used in detection of the land pre-pits.

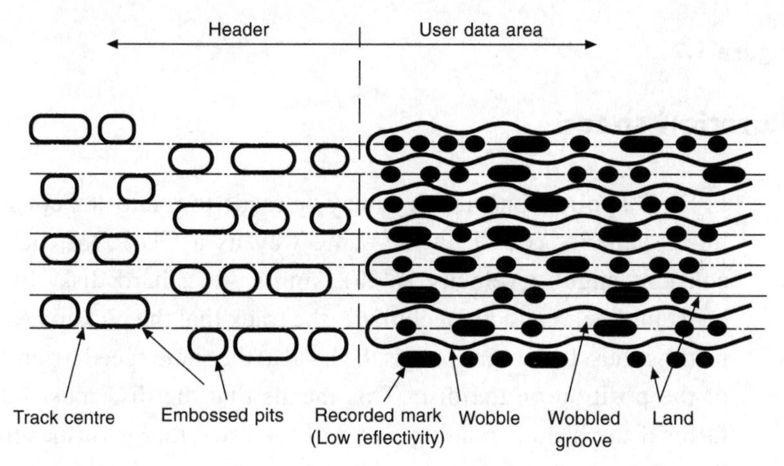

Figure 1.6

The pre-pits ensure high precision when writing the data, and provide the recording address and other information that are necessary for writing on a DVD.

After recording the disc Information Area, the playback region has exactly the same structure as that of a DVD-ROM disc, and the data format is also exactly the same. Closer to the centre of the disc than the Information Area is another region, called the R-Information Area, which is peculiar to DVD-R and DVD-RW discs. This area contains an area called the *PCA* (Power Calibration Area), which is used for laser power calibration, and an area called the *RMA* (Recording Management Area), which contains recording management information necessary for the recording device (Figure 1.7). This information is provided to prevent problems in playing these writable discs in ordinary players and drives.

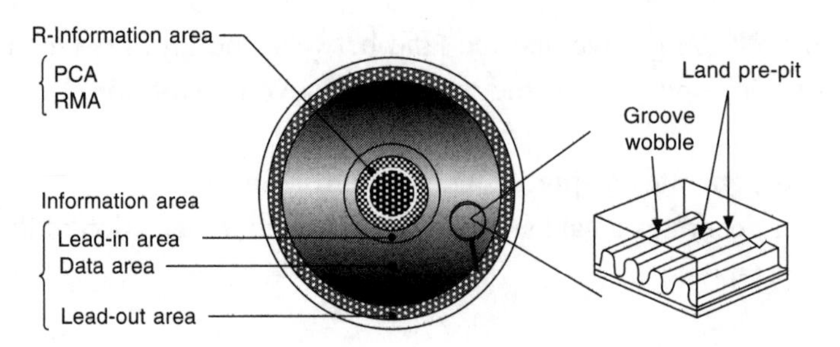

Figure 1.7

Rotation speed

In DVD applications, the disc may be made to rotate at a constant linear velocity (*CLV*), in the same way as a CD-ROM, or at a constant angular velocity (*CAV*), similar to a hard disk. In the constant linear velocity technique, the track that the pickup head is reading must be moving across the head at the same speed regardless of the position on the disc. This means that the disc must rotate faster if the head is reading a part of the track that is further from the outer circumference than if reading a part of the track that is nearer to the outer circumference. As the head follows the spiral track, the angular velocity of the disc must continuously change to keep the linear velocity constant. In the constant angular velocity technique the rotation of the disc is constant, resulting in the track moving faster across the head when the data being read are nearer to the outer circumference than when further from it. To keep the bit rate constant, pits nearer the outer circumference must be spread out compared with pits nearer the inner circumference. The result is a lower average data density compared with the CLV technique. Compared with CLV, CAV has the advantage of faster access to the various sectors on the disc surface. This is because when the pickup head moves to the required part of the disc, it does not have to wait for the angular velocity to change before reading the sector – as is the case with CLV.

CLV is used where the data on the disc are normally read sequentially, as in the case of a movie. Conversely, where the recorded data are not sequential and the pickup head has to move in search of the required sector, constant angular velocity with its fast access is employed.

While CAV provides fast access to the recorded data, it is inefficient in terms of its use of the available recording surface of the disc. To maintain the advantage of fast access while at the same time improving the efficient use of the disc surface, the zoned CLV disc layout is used. In the zoned CLV technique, the disc is divided into multiple concentric rings known as *zones* (Figure 1.8) in which the speed of rotation remains constant. The only change in angular velocity takes place when the pickup head moves from one zone to another. The user area of a DVD-RAM disc is divided into 24 (0–23) zones, with a lead-in and other areas preceding it and a lead-out area following it.

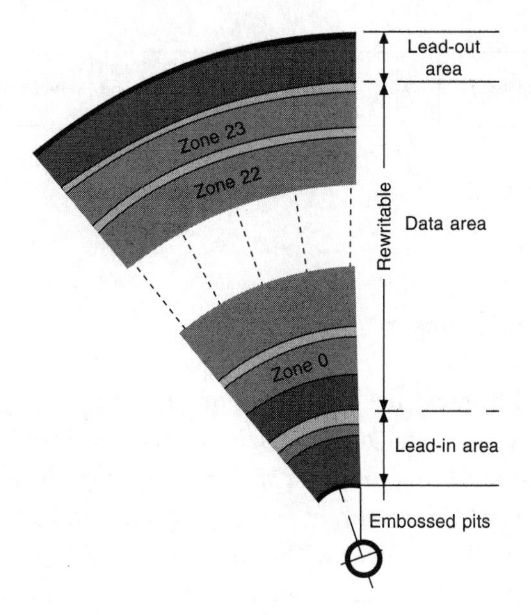

Figure 1.8

System overview

DVD technology employs a diverse range of advanced techniques, including optics, data compression and processing, electromechanical servo control and micro-processing, and digital and television technology. Optical technology is used to write data onto a disc and subsequently to retrieve it from the disc. The contents of a clip consist of video and audio information, which originate as analogue waveforms that require conversion to the digital format before compression and processing ready for burning onto a disc. Retrieving information from a disc involves reading the data, bit by bit, off a rotating disc, using a laser beam that must be positioned accurately to follow a few-micron spiral tracks on a rotating disc. The whole process of reading a DVD is controlled and all units programmed by a powerful microprocessor.

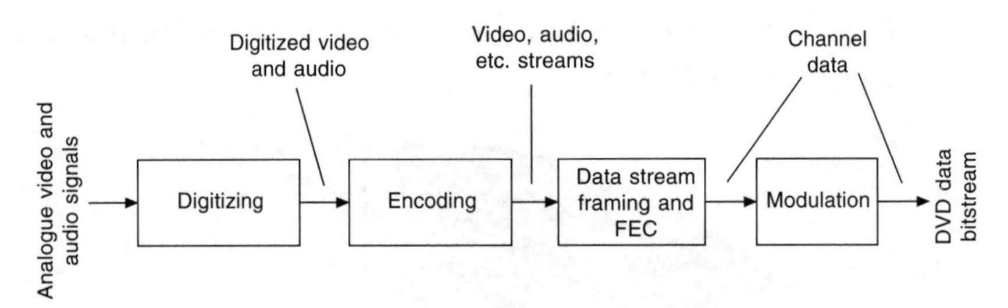

Figure 1.9

The encoding process

Figure 1.9 shows the basic units of a DVD encoding process. The analogue video and audio signals are first digitized before they are encoded to form a *program stream*. The program stream consists of audio and video *packets*, as well as control and other information that the DVD producer wishes to include. Each packet (known as a packetized elementary stream, *PES*) has a length of 2 KB (2048 bytes), and commences with a start code and is identified by a packet ID to distinguish it from others. The packets are manipulated to add appropriate error detection bits to form 2418-byte frames or sectors. The size of the frame is then doubled to 4836 bytes by a process known as eight-to-sixteen modulation to form the final bitstream, which is recorded onto the DVD – a process known as burning the disc.

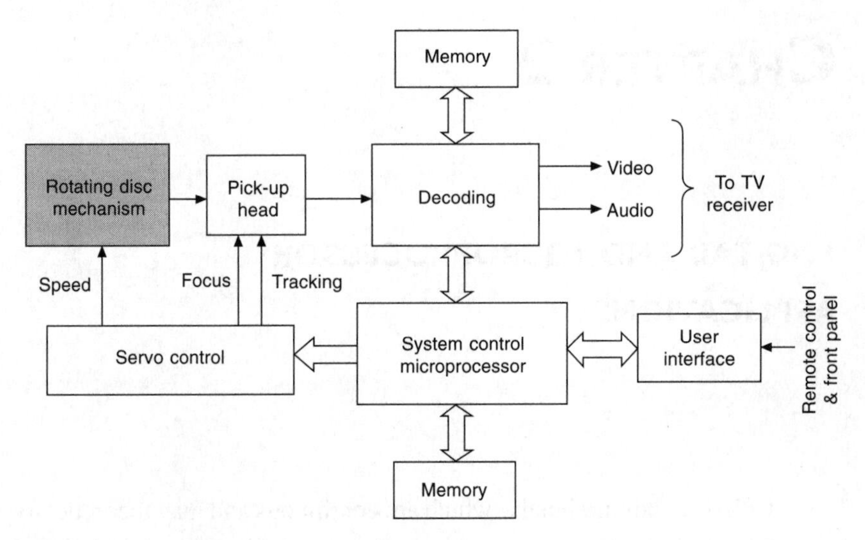

Figure 1.10

The playback process

During playback, the pick-up head reads the data on the disc, sector by sector (Figure 1.10). The speed of the disc, the position of the head and the beam focus are controlled by the servo control unit, which itself is controlled by the *system control* (*sys con*) microprocessor. The data from the pickup head are f

7ed into the decoding unit to reproduce the original audio and video signals. The process of decoding involves error correction and data decompression, both of which require the use of random access memory (RAM). Memory is also required by the sys con microprocessor to store the necessary processing routines, as well as various configurations and settings. The user accesses the player's facilities using a remote control or front panel switches, and signals from these are sent to the microprocessor for processing.

CHAPTER 2

DIGITAL AND MICROPROCESSOR APPLICATIONS

Unlike analogue signals, which are continuous and may theoretically take an infinite number of instantaneous values, a digital signal uses the binary system with two discrete values: logic '0' and logic '1'. A single binary digit, known as a *bit*, can thus have one of two values or states: a 0 or LOW, and a 1 or HIGH. The amount of information that can be exchanged using 1 bit only is very limited (ON or OFF, YES or NO, 0 or 1). Grouping a number of bits together to form what is known as a word may convey more information. A word using 2 bits (a 2-bit word) may be used to represent four different combinations ($2^2 = 4$), as shown in Table 2.1.

A 3-bit word doubles the quantity of information yet again, producing

Table 2.1 Combinations with 2 bits

Bit 1	*Bit 0*
0	0
0	1
1	0
1	1

eight different combinations ($2^3 = 8$; see Table 2.2) and so on. A 4-bit word, known as a *nibble*, provides sixteen different combinations ($2^4 = 8$), and an 8-bit word (known as a byte) provides $2^8 = 256$ different combinations.

Table 2.2 Combinations with 3 bits

Bit 2	*Bit 1*	*Bit 0*
0	0	0
0	0	1
0	1	0
0	1	1
1	0	0
1	0	1
1	1	0
1	1	1

Binary numbers

In the same way as denary (decimal) columns represent increasing powers of 10, binary columns represent increasing powers of 2; the rightmost bit is known as the least significant bit (*LSB*), and has a value of $2^0 = 1$. The next column has a value of $2^1 = 2$, the next $2^2 = 4$ and so on (Table 2.3). In any binary number, the leftmost bit is known as the most significant bit (*MSB*).

Table 2.3 Binary columns

	2^4	2^3	2^2	2^1	2^0	(16) (8) (4) (2) (1)
Example	1	0	0	1	1	$= 1 \times 16 + 0 \times 8 + 0 \times 4 + 1 \times 2 + 1 \times 1 = 16 + 2 + 1 = 19$
Example	0	1	1	0	1	$= 0 \times 16 + 1 \times 8 + 1 \times 4 + 0 \times 2 + 1 \times 1 = 8 + 4 + 1 = 13$

Logic levels

Each logic level (Figure 2.1) is represented by a voltage level. For *TTL* (transistor–transistor logic) technology, a logic *HIGH*

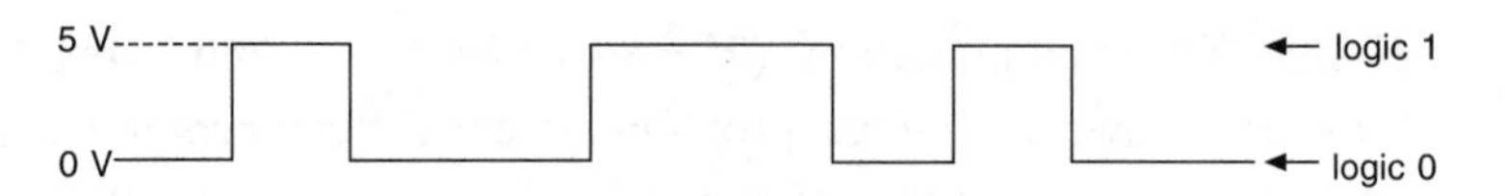

Figure 2.1
Logic levels for TTL

(logic 1) is represented by a high voltage of between 2.8 and 5 V and a logic *LOW* (logic 0) is represented by a low voltage of 0–0.4 V. For *CMOS* technology, logic levels are not represented by absolute voltages. A percentage of the DC supply voltage (V_{DD}) is used instead, whereby logic HIGH is represented by a voltage greater than 70 per cent and logic LOW by a voltage lower than 30 per cent of the V_{DD}. For CMOS devices, V_{DD} may be as low as 3 V and as high as 15 V – as opposed to TTL devices, which must have a very stable 5 V supply (represented by V_{CC}).

Hexadecimal

In order to avoid long strings of binary digits, hexadecimal notation is used. Hexadecimal numbers have a base of 16, and hence there are 16 distinct symbols:

0, 1, 2, 3, 4, 5, 6, 7, 8, 9, A, B, C, D, E, F

where A, B, C, D, E and F represent denary numbers 10, 11, 12, 13, 14 and 15 respectively. A single hexadecimal digit (Table 2.4) may thus represent a 4-bit binary number. An 8-bit binary number is represented by a two-digit hexadecimal number and a 12-bit binary by a three-digit hexadecimal number (Figure 2.2). A common way of distinguishing between binary and hexadecimal numbers is to terminate binary numbers with a B and hexadecimal numbers with an H. For instance, binary number 0010 is written 0010B, and hexadecimal number 2F as 2FH.

Table 2.4 Converting hexadecimal to binary

Hexadecimal	*Binary*
1	0000
2	0001
2	0010
3	0011
4	0100
5	0101
6	0110
7	0111
8	1000
9	1001
A	1010
B	1011
C	1100
D	1101
E	1110
F	1111

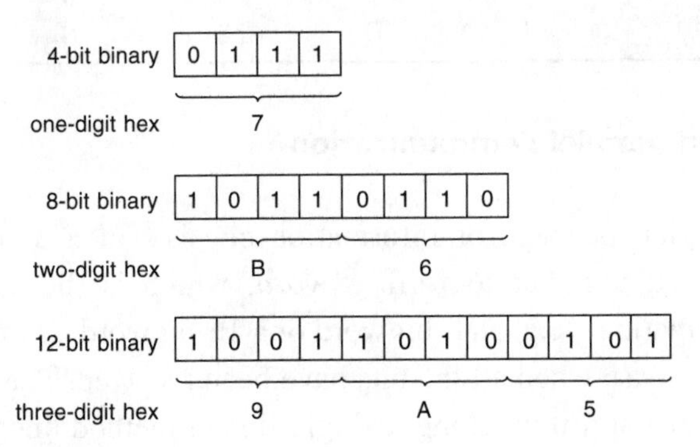

Figure 2.2
Binary-hex conversion

Logic gates

A logic gate is a device whose output is dependent on the instantaneous combination of its inputs. For instance, an *AND* gate will produce a logic 1 (High) output if, and only if, all of its inputs are High. The different types of gates, their symbols and truth tables are listed in Figure 2.3 and Table 2.5.

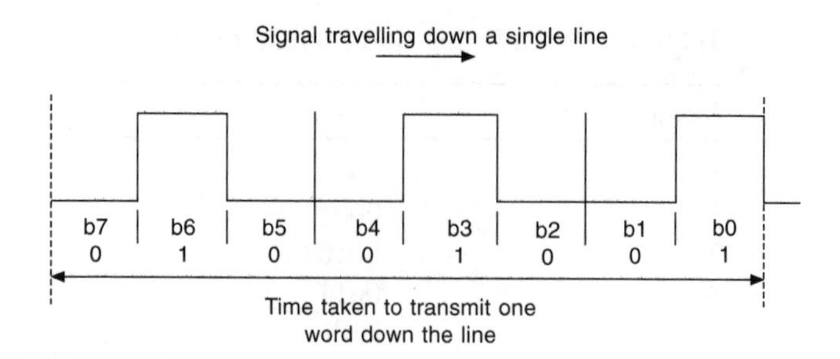

Figure 2.3
Serial transmission

Table 2.5 Truth tables for logic gates

Inputs		*Output function*				
A	*B*	*AND*	*NAND*	*OR*	*NOR*	*XOR*
0	0	0	1	0	1	0
0	1	0	1	1	0	1
1	0	0	1	1	0	1
1	1	1	0	1	0	0

Serial and parallel communication

A digital package of information consists of a number of bits grouped together to form a *word*, which is the basic unit of information – e.g. an 8-bit word or a 16-bit word. A word can only make sense when all the bits have been received. The bits may be sent one at a time along a single line, a method known as *serial* transmission (Figure 2.3), or they may be transmitted simultaneously, i.e. in *parallel* (Figure 2.4).

Shift registers

A shift register is a temporary store of data, which may then be sent out in a serial or parallel form. An 8-bit shift register is illustrated in Figure 2.5, in which serial data are clocked into the register, bit by bit. When the register is full, the data stored in it may then be

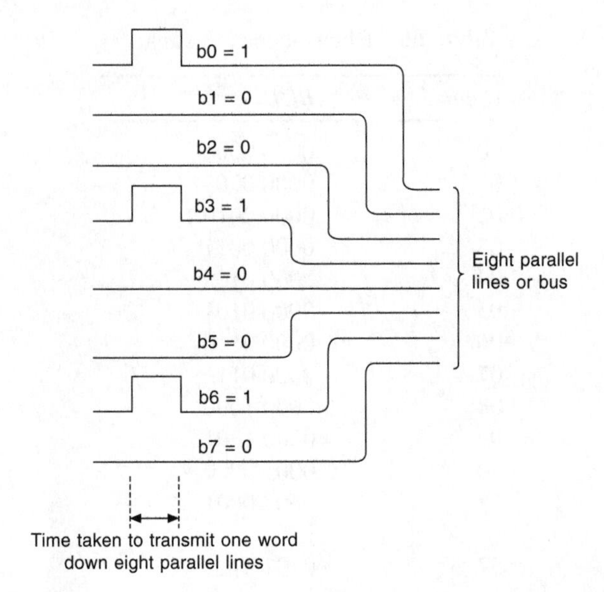

Figure 2.4
Parallel transmission

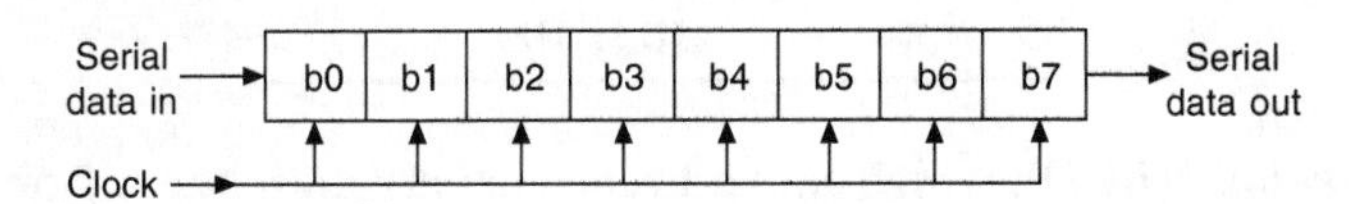

Figure 2.5
Shift register

clocked out serially, bit by bit. This type of shift register is known as serial-in serial-out (*SISO*) shift register. Three other possible arrangements are possible: serial-in parallel-out (*SIPO*), parallel-in serial-out (*PISO*), and parallel-in parallel-out (*PIPO*).

Digital codes

The conversion of a quality or a number into a digital format may be carried out using one of several codes. The natural binary code is listed in Table 2.6, in which the columns of, say, a 4-bit binary number represent progressively increasing powers of 2, giving a count of 0 to 15. Such a count is not very appropriate for denary applications. A more appropriate coding technique is *binary coded*

Table 2.6 Binary-coded decimal

Denary number	*BCD*
00	0000 0000
01	0000 0001
02	0000 0010
03	0000 0011
04	0000 0100
05	0000 0101
06	0000 0110
07	0000 0111
08	0000 1000
09	0000 1001
10	0001 0000
11	0001 0001
⋮	⋮
57	0101 0111
⋮	⋮
83	1000 0011
⋮	⋮
99	1001 1001

decimal (*BCD*), which converts each denary digit into a 4-bit binary number. A 2 digit BCD number will thus result in two groups of 4-bit binary numbers (8 bits in total).

Another popular coding technique is the Gray code. This ensures that only 1 binary bit changes state, as the denary number is progressively incremented (Table 2.7). The Gray code avoids the problems of spurious transitional codes associated with the BCD technique.

Multiplexing

Communication invariably involves transmitting several packages belonging to different programs along the same communication media, such as cable, satellite or terrestrial links. There are two methods by which a communication medium is shared between several packages: *broadband*, using *frequency-division multiplexing*

Table 2.7 Gray coding

Denary	*Binary code*	*Gray code*
0	0000	0000
1	0001	0001
2	0010	0011
3	0011	0010
4	0100	0110
5	0101	0111
6	0110	0101
7	0111	0100
8	1000	1100
9	1001	1101
10	1010	1111
11	1011	1110
12	1100	1010
13	1101	1011
14	1110	1001
15	1111	1000

(FDM), and *baseband*, using *time-division multiplexing* (*TDM*). The former frequency-division technique involves dividing the available bandwidth into several channels, and each channel is then allocated to carry the packages belonging to a single program. The packages are thus transmitted simultaneously. In time-division multiplexing, on the other hand, the packages are transmitted sequentially. Each package is allocated a time slot during which the whole of the bandwidth of the medium is made available to it. At the receiving end, the transmitted packages are demultiplexed to reconstruct the original programs (see Figure 2.6).

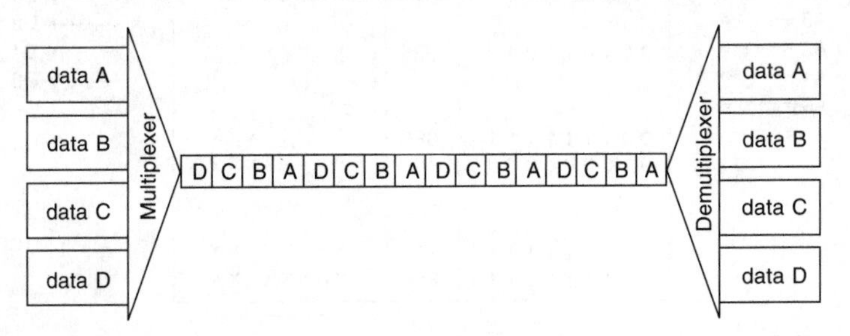

Figure 2.6
Multiplexing

Memory chips

Memory chips are data storage devices where information in the form of software programs or text may be saved.

A memory chip contains a number of locations, each of which stores one or more bits of data. This determines what is known as its bit width. Each location is identified by a unique *address*. Figure 2.7 shows the basic organization of a memory chip. Each location stores 8 bits of data, known as a byte, and is given a unique binary address. An address with 3 bits (A0, A1 and A2) provides a maximum of $2^3 = 8$ memory locations; a 10-bit (A0–A9) address has $2^{10} = 1024$ (or 1K) memory locations (Figure 2.7). The address lines are grouped together to form an address bus, and by placing an address on the address bus, any one of the 1024 locations may be selected.

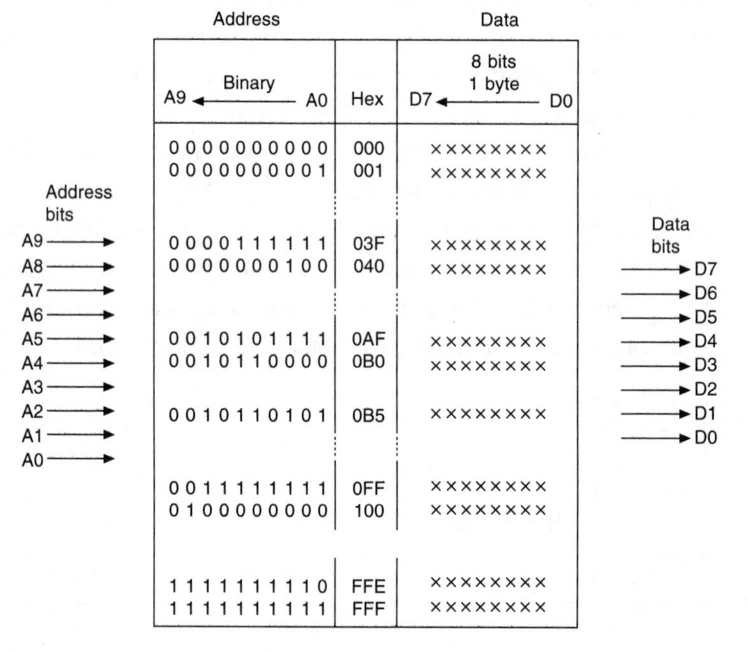

Figure 2.7
Basic organization of a memory chip

For instance, if A9, A8, . . ., A1, A0 are set to logic levels 0, 0, 1, 0, 1, 1, 0, 1, 0, 1, then the location with binary address 0010110101 (0B5 in hexadecimal) will be selected. For simplicity, addresses are normally stated in hexadecimal. Once a memory location is chosen, pins D0 to D7 on the chip provide access to the eight memory cells in that location. Data are also normally stated in hexadecimal.

Memory chips have two main properties: *storage capacity* (size) and *access time* (speed).

Storage capacity

The storage capacity of a memory chip is the product of the number of locations by the data bit width. For example, a chip with 512 locations and a 2-bit data width has a memory size of

$$512 \times 2 = 1024 \text{ bits}$$

Since the standard unit of data is a byte (8 bits), storage capacity is normally given as

$$1024/8 = 128 \text{ bytes}$$

The number of locations is determined by the number of address pins that are provided. For example, a chip with 10 address lines has $2^{10} = 1024$ (or 1K) locations. Given an 8-bit data width, a 10-bit address chip has a memory size of

$$2^{10} \times 8 = 1024 \times 8 = 1\text{K} \times 1 \text{ byte} = 1\text{K byte, or } 1\text{KB}$$

A single chip is usually insufficient to provide the memory requirements of a computer, and a number of chips are therefore connected in parallel to form what is known as a memory bank.

Access time

Access time is the speed with which a location within the memory chip can be made available to the data bus. It is defined as the time interval between the instant that an address is sent to the memory chip, and the instant that the data stored into the location appear on the data bus. Access time, given in nanoseconds (ns), ranges from very fast (10 ns) to relatively slow (200 ns).

Types of memory chips

Random access memory

Random access memory (*RAM*) is a memory chip that the user may read from or write into, and hence it is also known as read/write memory. RAM chips are known as '*volatile*', because their contents are lost when power is switched off; however, one method of preserving the stored data is to employ back-up batteries to maintain the DC supply voltage to the chip when the mains supply is removed.

Locations may be accessed at random by placing the address of the selected location onto the address lines. The pin-out requirements of a RAM chip are shown in Figure 2.8. Apart from the address

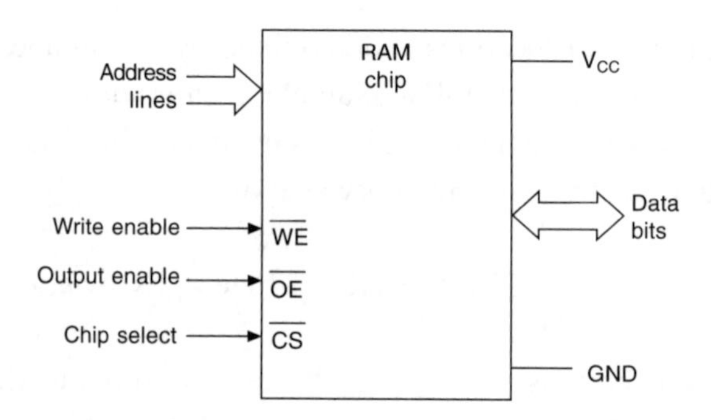

Figure 2.8
RAM chip standard pin-out

lines and the data bits there are three control lines, all of which are *active LOW*: Write Enable (WE), which goes LOW when the CPU wishes to write new data into the selected location; Output Enable (OE), which goes LOW when the processor wishes to read the content of the location; and Chip Select (CS), which is driven LOW when the selected location falls within the range address assigned for the chip.

There are two major categories of RAMs: Dynamic RAM (DRAM) and Static RAM (SRAM). *Dynamic RAMs* store information in the form of a charge on a capacitor. However, owing to leakage the charge is lost, and has to be restored at frequent intervals of between 2 and 4 ms – a process known as '*refreshing*' the cells. Dynamic RAMs have the advantages of higher component density (and hence small size) and very low power consumption. *Static RAM* devices employ flip-flops (electronic switching devices) as the basic cell, and hence require no refreshing. They will hold data as long as DC power continues to be applied to the device. SRAMs are very fast, with an access time of 20 ns or less (compared with 60 ns for DRAMs). However, they are more expensive and larger in size than the dynamic type, which inhibits their use as the main memory store of a computerized system.

Owing to their low cost and high component density, DRAM devices are used to provide the bulk of system memory of few megabytes. The number of address pins required to accommodate this size of memory becomes physically inhibitive for manufacturing purposes, and to overcome this problem address multiplexing is employed. The multiplexer receives the full address of, say, 20 bits from the address bus, which is then fed to the memory chip in two stages. First A0–A9 is fed to the address pins on the IC, followed A10–A19, which are fed to the same IC pins. Two special control signals, *column address strobe* (*CAS*) and *row address strobe* (*RAS*), are provided to root the two halves to two internal latches. The full address is then held within the IC long enough to access the data in the selected location. A typical pin-out is shown in Figure 2.9.

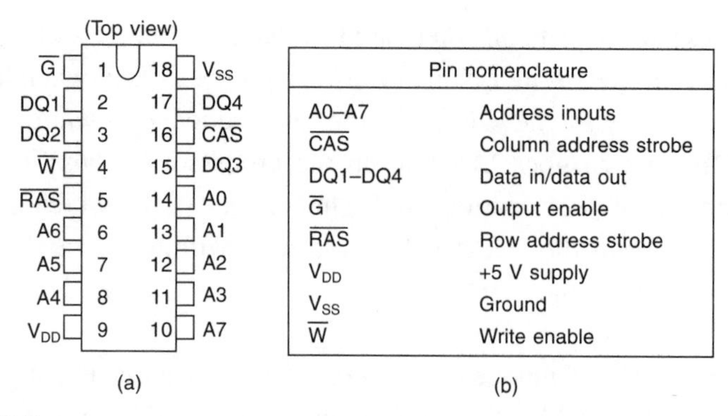

Figure 2.9
Dynamic RAM: typical pin-out

Back-up battery

One method of preserving the stored data in a normal RAM is to employ back-up batteries to maintain the DC supply voltage to the ship when the mains supply is removed.

Read only memory

Read only memory (ROM) chips are '*non-volatile*', in that they retain their data irrespective of the power supply. Compared with RAM, ROM devices are slow, with a typical access time in excess of 200 ns. This makes them unsuitable for applications that require fast memory access – such as video applications. There are several types of ROM devices.

Mask ROM is a non-volatile memory used for storing data permanently. The stored data can only be read by the user, and no new data can be written into the device. ROM is programmed by the manufacturer in accordance with predetermined specifications. Once entered, the data cannot subsequently be altered.

PROM (programmable read only memory) devices fulfil the same basic function as ROM chips, except that they may be programmed by the user – a process known as '*blowing*' the chip. Once programmed, PROMs cannot be altered.

EPROM (erasable programmable read only memory) chips overcome this by allowing the user to delete or erase the stored data and thus change the program. The stored program in an EPROM may be erased by exposing the memory cells to ultraviolet light through a 'window' on the IC package. This process takes 20 to 30 minutes, at the end of which the IC is in a 'blank state' ready to be reprogrammed.

EEPROMs (electrically erasable programmable read only memory) chips can be programmed and erased while still connected to the circuit, by the application of suitable electrical signals. Furthermore, individual locations may be erased and programmed without interfering with the rest of the data pattern. As a result of overwriting location, EEPROM has a comparatively short lifespan.

Flash RAM is an advance on the EEPROM. Once again all locations may be erased and reprogrammed, but this time using normal voltages that are available in a computerized system. However, Flash RAM continues to suffer from a short lifespan and a long access time of between 60 and 150 ns.

Table 2.8 provides a comparison of the types of memory chips.

Table 2.8 Memory chip comparisons

	DRAM	*SDRAM*	*SRAM*	*ROM*	*EPROM*	*EEPROM*	*Flash*
Capacity	Mbytes	Mbytes	kbytes	kbytes	kbytes	kbytes	kbytes
Speed (ns)	60	10	2–20	200	400	400	80
Lifespan	long	long	long	long	short	short	short

General purpose microprocessor systems

The architecture of a microprocessor-based system is shown in Figure 2.10. It consists of the following component parts:

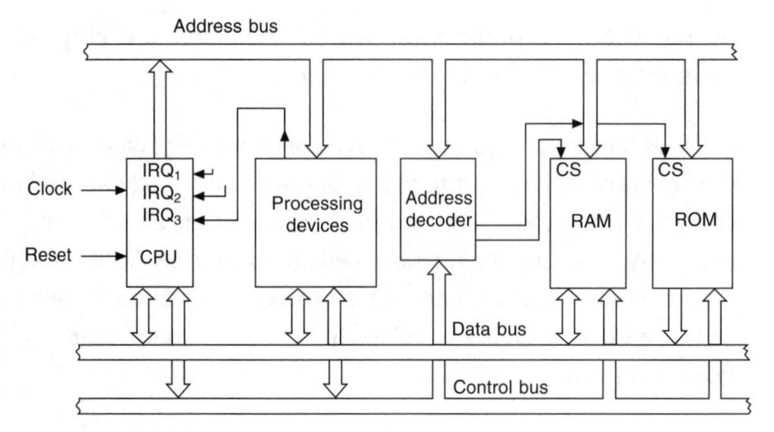

Figure 2.10
Components of a microprocessor-based system

- Central processing unit (CPU)
- Memory chips (RAM and ROM)
- Address decoder chip
- Processing devices
- Bus structure.

The central processor unit

The CPU is usually a single *VLSI* (very large scale integration) device chip containing all the necessary circuitry to interpret and execute program instructions such as data manipulation, logic and arithmetic operations, and timing and control of the system. The capacity or size of a microprocessor chip is determined by the number of the data bits it can handle – a 16-bit chip has a 16-bit data width; a 32-bit processor has a 32-bit data width and so on. Eight-bit and 16-bit processors are generally employed as dedicated controllers in industrial applications and in domestic appliances such as robots, washing machines and TV receivers.

Microprocessors differ in the speed at which they execute instructions, and CPU speeds are indicated by the frequency of the system clock in MHz. While the bit width or size determines the quantity of information that may be transferred in any one cycle, the speed determines the number of transactions that may be executed per second.

Memory chips

Microprocessor systems require a certain amount of data storage capacity where programs such as start-up routines and other processing software reside. DRAM and ROM are two types of memory chips that are normally used to provide the necessary memory storage space. Other types, such as SDRAM, SRAM, PROM, EPROM, EEPROM or Flash, may also be used.

Address decoder chip

The *address decoder* receives a group of address lines and, depending on their combination, enables one of its outputs – normally by taking it to logic LOW. If this line is connected to the *chip select* (CS) pin of a memory chip, then that memory chip will be enabled (i.e. selected). With two address lines, four (2^2) outputs are available; with three address lines, eight (2^3) outputs are available, and so on. Figure 2.11 shows a typical two to four address decoder with its truth table.

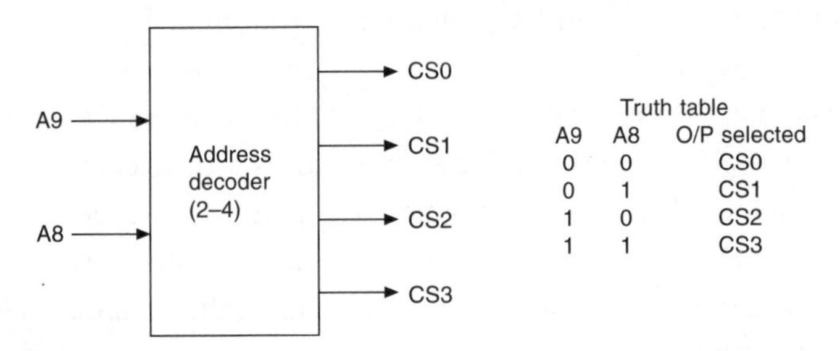

Truth table

A9	A8	O/P selected
0	0	CS0
0	1	CS1
1	0	CS2
1	1	CS3

Figure 2.11
Address decoder and its truth table

Processing devices

DVD players, like almost all consumer and industrial electronic equipment, employ one or more processing devices to carry out such functions as programming and control of the system and system devices. Four types of processor chips may be used in a DVD player:

1. General purpose microprocessors for general system programming and control
2. Dedicated microprocessors such as video decoders or RF processors
3. Microcontrollers
4. System-on-chip processing chips.

Three main technologies are employed in the fabrication of integrated circuits: TTL, CMOS and NMOS. The latter two types are normally used because of their high component density.

The bus structure

The hardware elements described above are interconnected by a bus structure consisting of three types of buses: *address, data* and *control* (see Figure 2.10). The address and data buses provide a parallel highway along which multi-bit addresses and data travel from one unit to another. The control bus incorporates the lines that carry the control signals to and from the CPU.

The data bus is used to transfer data between the CPU and other elements in the system. The address bus is used to carry the address of memory locations from which data may be retrieved (i.e. *READ*) from memory devices, or where data may be stored (i.e. *WRITE*). It is also used to address other elements in the system, such as the input/output ports. The control bus carries the control signals of the CPU, such as the clock, RESET, READ (RD) and WRITE (WR).

Control signals

The number and type of control signals depend on the microprocessor used and the design of the system. Control signals are normally active LOW – i.e. active when at logic 0. Active LOW signals are signified by a bar or by a small circle at the chip pin-out, as illustrated in Figure 2.12.

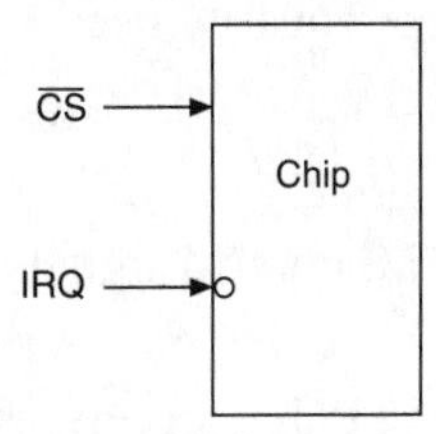

Figure 2.12

The main control signals of a CPU are described below.

The clock

A clock pulse (Figure 2.13) is an essential requirement for the operation of the processor. The clock control signal synchronizes the movement of the data around the various elements of the system and determines the speed of operation; without it the system comes to a halt. A crystal-controlled oscillator is used to provide accurate

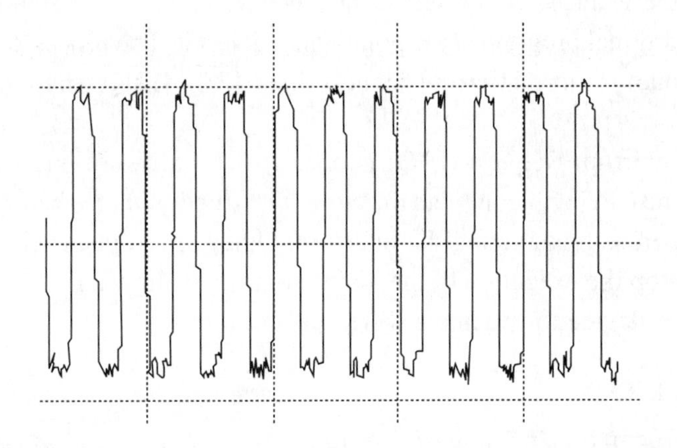

Figure 2.13

and stable timing clock pulses. Clock frequencies range from the relatively slow (10 MHz) to the faster 100 MHz and over. In personal computers, the system processor clock frequency may be as high as 2000 MHz. Stable clock frequency is essential, and a small drift may cause the processor to malfunction. Clock pulses may be monitored by using a logic probe. However, to test for the accuracy of the clock frequency, an oscilloscope with appropriate bandwidth or a frequency counter must be used.

READ and WRITE

The READ and WRITE control signals determine the direction of data transfer to or from the microprocessor chip. In a READ operation (when the CPU is receiving data from memory) the READ line is active, allowing data to be transferred to the CPU. In a WRITE operation (when the CPU is sending data to memory) the WRITE line is active, enabling data transfer from the CPU to memory.

Interrupts

When a peripheral device such as a channel decoder or a transport demultiplexer needs attention, a hardware interrupt request signal (IRQ) is sent to the CPU. When such a signal is received, the main program is interrupted temporarily to allow the CPU to deal with the request. After servicing the device, the CPU returns to the original program at the point where it left it. The processor provides one or more interrupt request lines $\overline{IRQ}_1$–$\overline{IRQ}_3$ in Figure 2.10.

Interrupt Request (IRQ) is one type of hardware interrupt where the CPU will complete the current instruction that is being executed before recognizing the interrupt. Other interrupts, such as HALT, stop the execution of the main program to allow an external source or device to execute a different program.

RESET

$\overline{\text{RESET}}$ ($\overline{\text{RES}}$ or $\overline{\text{RST}}$) is a type of interrupt that overrides all other interrupts. The RESET input pin is normally held HIGH. If it is taken LOW, it immediately stops the CPU program and the processor is reset. To restart the microprocessor operations, the RESET pin must be taken HIGH again.

Figure 2.14 shows a press-button manual reset circuit. When power is switched on and a 5 V DC supply established, capacitor C1 charges up through the 1 kΩ resistor. The RC network provides a delay, which is necessary to avoid switch bounce. It is also necessary to ensure that all DC voltages have reached a steady state before the processor is initialized. (A J–K flip-flop is sometimes used in place of the simple RC circuit). After a short time, which is determined by time constant C1R1 (0.1 s), the reset pin goes to logic HIGH. When that happens, the microprocessor immediately commences an initialization sequence. This sequence consists of directing the CPU to the memory location where the start-up program of the system is stored. The RESET pin is held at logic HIGH via the 1 kΩ resistor. Two TTL low-power Schottky (LS) inverter/ buffers (74LS04) are used to ensure that the correct logic levels are established. This simple circuit may be suitable for a system with a single programmable chip, such as a CPU; however, where a number of programmable chips are involved (as is the case with a DVD player), a more complex resetting and initializing arrangement is necessary.

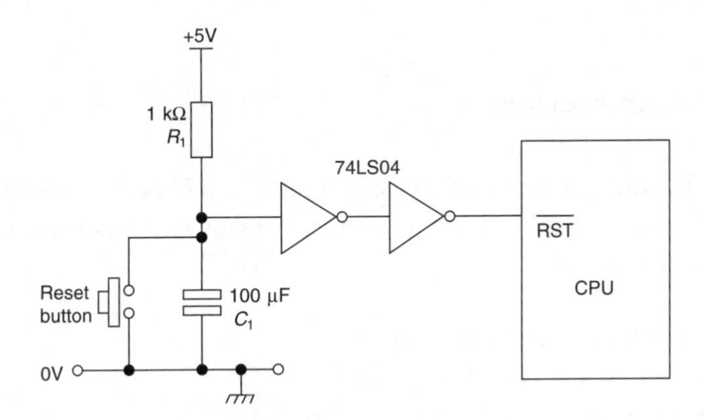

Figure 2.14
Simple reset circuit

Resetting and initializing

All programmable chips, including the CPU and power supply control chips, must have their registers set to an initial valve, which determines the start-up parameters of the chips. This is known as initialization. Initialization involves taking the REST control line from GND level to +5 V. The simple RC network shown in Figure 2.14 produces a reset pulse known as Power On Reset (POR), which is used to reset and initialize all programmable chips. The pulse is made active only after the various DC voltages have reached a steady and stable level, and this normally requires a delay of 20–60 ms.

The main pin-out of a basic microprocessor chip is illustrated in Figure 2.15a. A typical pin-out of a microprocessor used in a DVD player is shown in Figure 2.15b, in which pin 1 is indicated by a dot. The chip supports a 24-bit address bus (pins 58–81), and a 16-bit data bus (pins 40–55). RESET is on pin 34, the clock is on pin 26, and pins 1–7 are allocated for use for DMA request and acknowledge as indicated. The chip supports five interrupts (INT_0–INT_4) on pins 98–100 and pins 94/95, and two sets of bidirectional three-line serial communications – each incorporating Receive (Rx), transmit (Tx), and a clock line (pins 82–84 and pins 86–88). Memory control lines include READ pin 30, WRITE pin 18 and CAS/RAS pins 16–21.

CPU architecture

The microprocessor chip has a complex architecture, which varies from one manufacturer to another. However, it has the following common units:

- Arithmetic and logic unit
- Timing and control logic unit
- Accumulator and other registers
- Instruction decoder
- Internal bus.

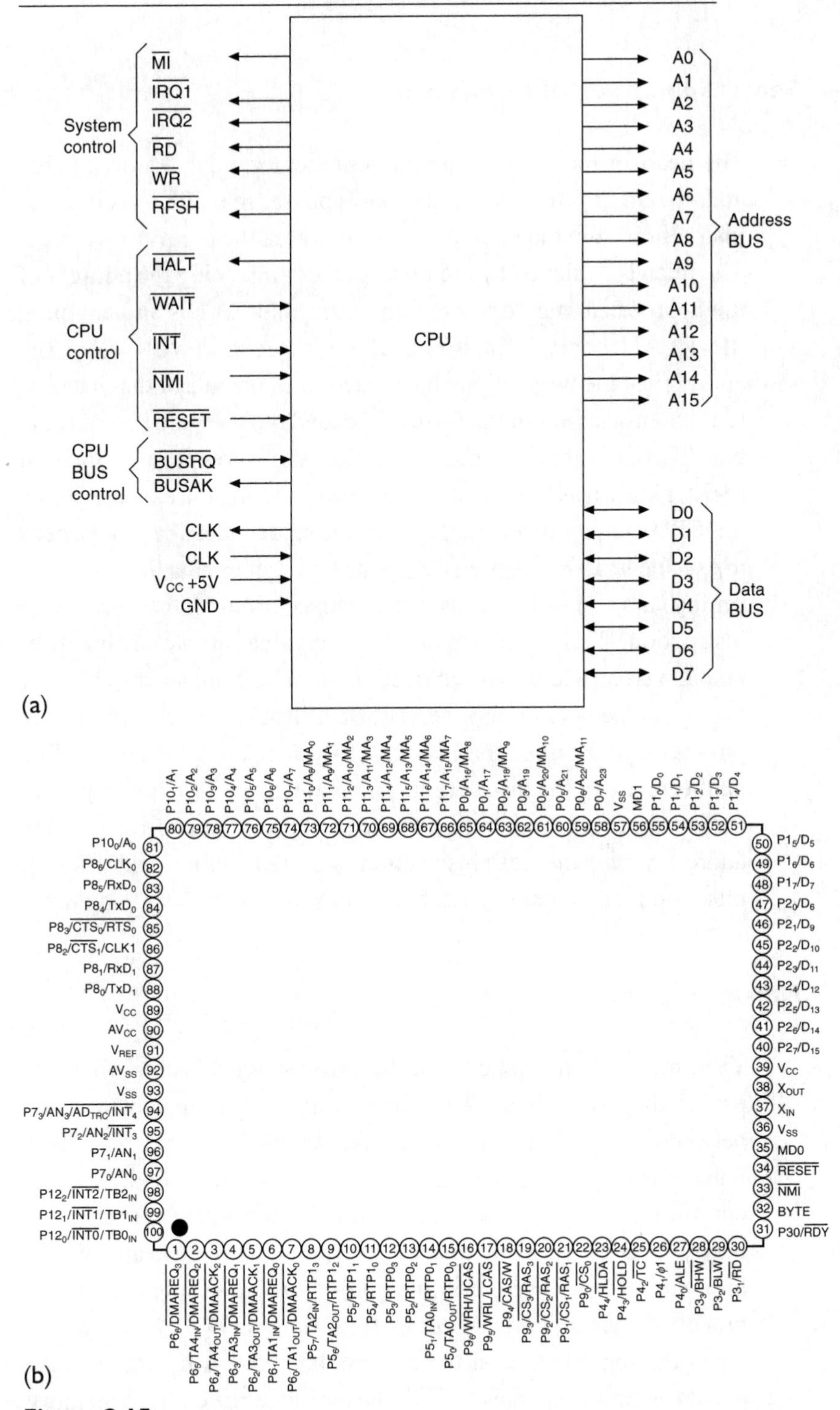

Figure 2.15
CPU pin-out

General operation of the system

The heart of the system, the microprocessor, operates on a Fetch and *Execute* cycle. During the fetch phase, the CPU receives the instruction from the memory location where the program is stored. The fetch is achieved by the microprocessor placing the address of the appropriate memory location on the address bus and enabling the READ control line. The address decoder will select the appropriate memory chip, which places the contents of that location (i.e. the instruction in the form of a coded binary word) on the data bus. The CPU receives the instruction and stores it in an internal register known as the *instruction register*. During the execute phase, the CPU, having received the instruction, decodes it and proceeds to execute it. This is carried out by the CPU generating the necessary timing and control signals for the execution of that particular instruction. The execute phase may involve simple arithmetical operation (e.g. addition or subtraction) or more complex data transfer to or from memory or peripheral devices. Both the fetch and execute phases may take more than one clock cycle to complete, depending on the nature of the instruction. When the instruction is completed, the microprocessor then places the next program address (i.e. the address where the next instruction is stored) on the address bus, thus commencing another fetch and execute cycle, and so on.

Timing diagram

The time relationship between the various signals is known as the timing diagram. Figure 2.16 shows a typical timing relationship between the clock signal, address, data and READ lines. The clock pulse ensures precise synchronization of address, data and RD control signals. The bits corresponding to the address from which they are to be read are placed on the address bus, and this is followed by the RD control line going LOW to enable the reading process. After a short period of time (one or two clock cycles), when the logic levels on the address bus have had time to settle and the address becomes 'valid', the data is retrieved from memory.

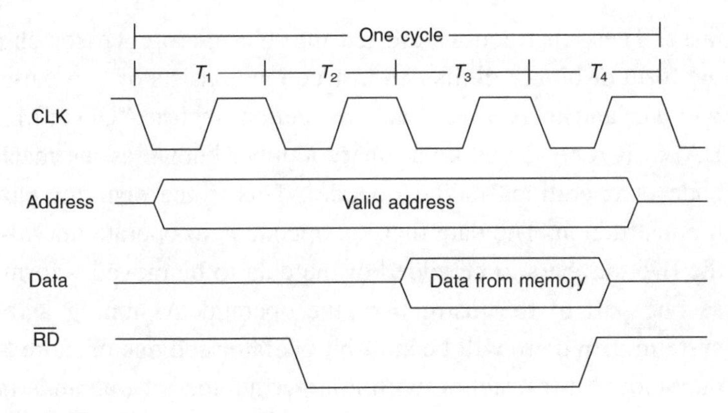

Figure 2.16
Read cycle

The process is then repeated for the next read cycle and so on. Figure 2.17 shows the timing diagram for a WRITE cycle.

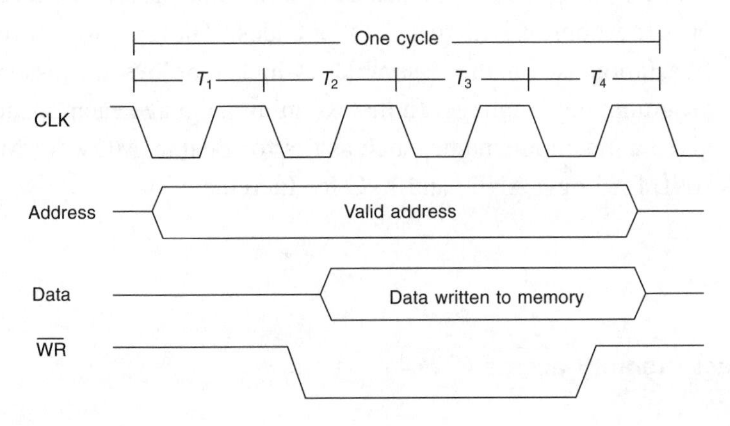

Figure 2.17
Write cycle

The instruction set

The microprocessor performs its tasks in a predetermined sequence known as the *program*. The program is a series of instructions, which break down each operation into a number of individual tasks. These instructions are fed into the microprocessor chip in the form of binary digits. An instruction consists of two parts: an *operator* and an *operand*. Each instruction, such as ADD or MOVE DATA, is represented by a binary number known as the machine code or operational code (opcode). This is the operator part of the instruction. The data that the opcode is to operate upon – i.e. the two numbers to be added or the data to be moved – form the second part of the instruction, the operand. Assuming an 8-bit system, then there will be an 8-bit operator and one or more 8-bit operands. An instruction with a larger number of operands takes longer to complete than one with fewer operands. Each make of microprocessor has its own set of machine codes, known as the instruction set.

Writing programs directly in machine code is a very lengthy and tedious process. Normally programs are written in a language that uses normal alphabetical letters and words, and this is then translated into the appropriate series of opcodes. The simplest form of translation is via the assembler, which employs the assembly programming language. In the assembly language each opcode is given a mnemonic name, such as EN for Enable, MOV for Move, ANL for Logic AND, and INC for Increment.

Direct memory access (DMA)

The vast majority of computer operations involve the transfer of data between different units of the system, and the processor itself normally carries this out. Where a large amount of data is to be transferred (e.g. complex graphic or video applications) a faster method, known as direct memory access (DMA), is normally used. Here, a controller (DMAC) takes control of the system for the duration of the transfer. Once the data transfer has been completed, the DMAC hands over control back to the CPU.

Microprocessor types

There are two basic types of microprocessors: *CISC* (complex instruction set code) and the faster *RISC* (reduced instruction set code). RISC processors are capable of performing fast mathematical operations using fewer or 'reduced' number of instructions. Examples of CISC processors include all the Intel 80XXX and Pentium series. Examples of RISC processors include the ARM and OAK.

Single-chip computers

The elements of a microprocessor system may be incorporated on a single chip to form what is known as a single-chip computer (Figure 2.18). They are used to control other devices, such as digital-to-analogue converters, decoders, and digital signal processors (DSPs).

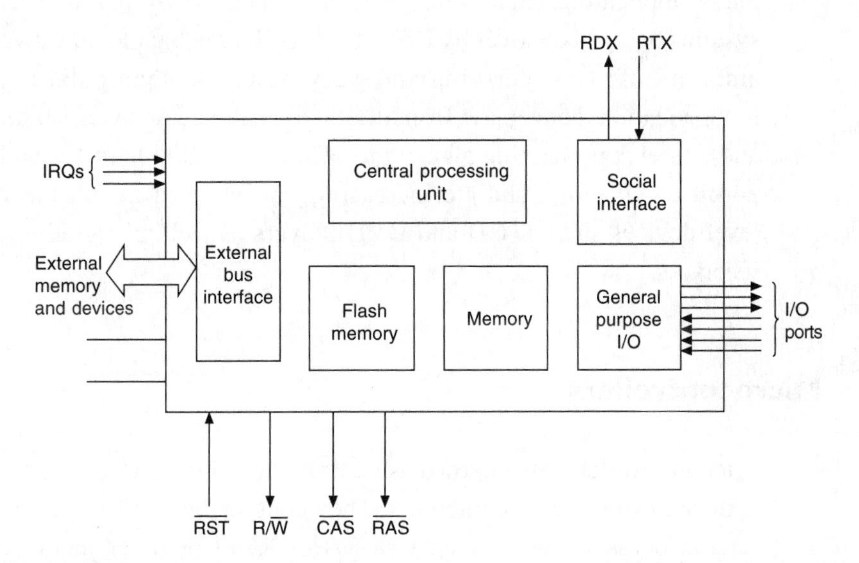

Figure 2.18
Computer parts of a single-chip computer

Dedicated processors

Dedicated processors are processors that are dedicated to one particular task, such as servo control and video/audio decoding. They are programmed and controlled by the resident system processor via the address/data bus and control signals, or by a serial bus, or by both. Processing chips have their own individual chip clock, which determines the processing speed. They normally require their own dedicated memory store, which is accessed by a dedicated address/data bus. The chip clock is distinct from the system clock, which provides the necessary system synchronization.

Digital signal processors

Recent advances in microprocessor technology have resulted in the development of very fast digital signal processor chips. The introduction of faster and more compact silicon chips with the increasing sophistication of processing technology has resulted in many applications that were previously handled by board level systems being taken over by DSP chips. DSPs can handle intensive amounts of data, carrying out very fast data manipulation, multiplication, conversion from digital to analogue and *vice versa*, and complex processing algorithms. They are available in 16- and 24-bit architecture, and are increasingly being used in consumer level devices such as CD and DVD players as well as digital TV receivers.

Microcontrollers

Microcontrollers, also known as central control units (CCUs), are dedicated single-chip computers. They contain the elements of the microprocessor itself, as well as RAM, ROM or other memory devices and a number of input/output ports. A variety of microcontrollers are available from various manufacturers (for example the Intel 8048/49 and 8051 series, Motorola 6805 and 146805, Texas TMS1000 and Ziloc Z80 series) for use as dedicated computer systems in such applications as car engines, washing machines, VCRs and TV receivers. The difference between one type of microcontroller and another lies in the type and size of memory, the instruction set, operating speed, the number of available input and output lines, and the data width (i.e. 4, 8 or 16 bits). In the majority of cases microcontrollers have their program stored permanently into an internal ROM at the manufacturing stage, a process known as mask programming. Some chips have an external EPROM available for user programming.

The serial control bus

A serial bus is invariably used by microprocessors and microcontrollers to control other units in the system. While manufacturers may use their own proprietary serial control bus systems, there are standard systems – namely the two-line inter IC, commonly known as IIC or I^2C bus, and the three-line intermetall (IM) bus.

The *I^2C bus* has two bidirectional lines, a *serial clock* (*SCL*) and *serial data* (*SDA*). Any unit connected to the bus may send and receive data. Data are transmitted in 8-bit words or bytes, as shown in Figure 2.19. The first byte contains the 7-bit address of the device for which the information is intended, while the eighth bit is a read/write bit to signify whether the data is required from, or being sent to, the device. A number of data bytes follow, the total number in a message depending on the nature of information being transferred. Each data byte is terminated by an acknowledge (ACK) bit. Like all other bits, the ACK bit has a related clock pulse on the clock line.

The first byte of any data transfer is preceded by a start condition and terminated by a stop condition. To ensure that two devices do not use the bus simultaneously, an arbitration logic system is used.

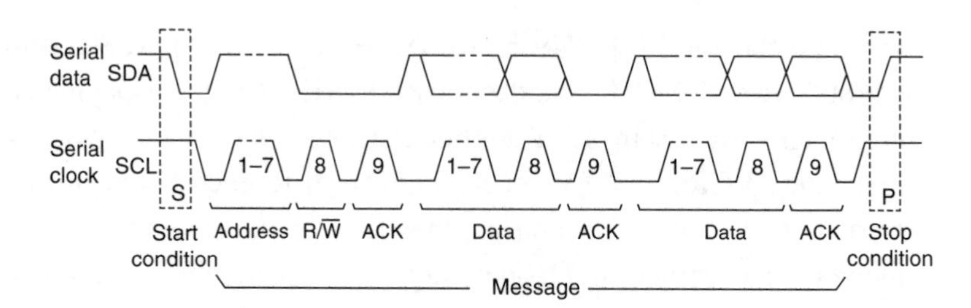

Figure 2.19
Two-line I^{2}C serial control bus

The clock, which operates only when data are transferred, has a variable speed. Data may then be sent at a slow or a fast rate of up to 100 kbit/s.

The *Intermetall bus* has three lines; *Ident* (I), *Clock* (C) and *Data* (D). Both the Ident and the Clock lines are unidirectional between the microcontroller and the other peripheral devices, whereas the Data line is bidirectional. The start of transmission is indicated by the Ident line going LOW (Figure 2.20), and an 8-bit address is sent along the Data line. At the end of eight clock cycles the Ident line goes HIGH, indicating the start of data transmission. Data are then transmitted along the D line for 8 (or 16) clock cycles for an 8-bit (or 16-bit) data word, at the end of which the I line goes LOW again, indicating the end of data transmission.

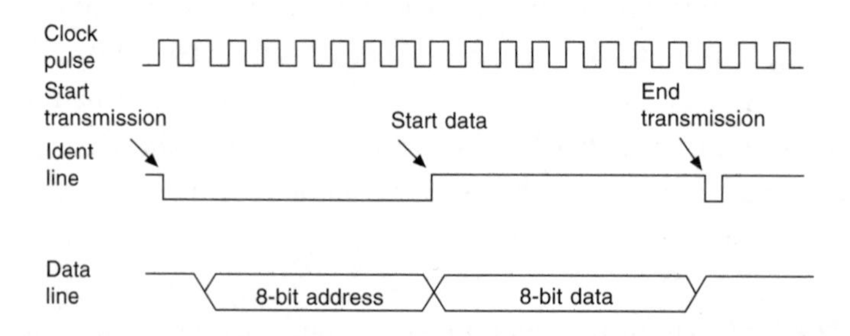

Figure 2.20
Three-line intermetall serial control bus

Both busses may be used in a single receiver to provide connections with different sections of the receiver. A number of peripheral chips are available on the market, including tuner interfaces, channels decoders, EEPROMs, ADCs, and a variety of digital signal processors (DSPs) for operation with I^2C or IM buses.

System-on-a-chip

Evolution in chip technology has vastly increased the density of integrated circuits and their functionality. Changing from a 0.5-μm integrated circuit technology to 0.18 μm has resulted in a vast unused space known as '*white space*' (Figure 2.21), which may be utilized for extra functions. This space is used to perform various processes (analogue or digital), as well as providing DRAM, SRAM, EPROM or Flash embedded memories. Merging different processes in this way produces a single universal processor known as a system-on-a-chip (SoC) processor.

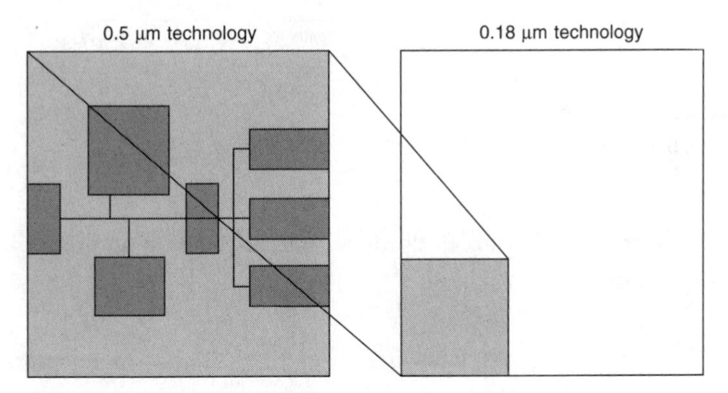

Figure 2.21
New IC technology has led to vast amounts of unused 'white' space

System-on-a-chip is the latest advance in chip technology, and is expected rapidly to replace ASIC (application specific integrated circuit) technology in the next few years. SoC combines the core of a microprocessor with embedded memory space (DRAM, SRAM or Flash), I/O ports, serial UART and an external bus interface

(Figure 2.22). A special interface for testing purposes is usually provided, designed for use at the testing and development stage. Apart from reducing the number of chips in a chip set, SoC reduces the power requirements of the system. The core is based on a powerful RISC processor chip, such as from the ARM or OAK families. SoC processors carry out two functions simultaneously:

1. They program and control external devices (such as video/audio decoders and servo control processors) using their powerful CPU core
2. They carry out a specific complex processing operation, such as transport demultiplexing in a digital television decoder.

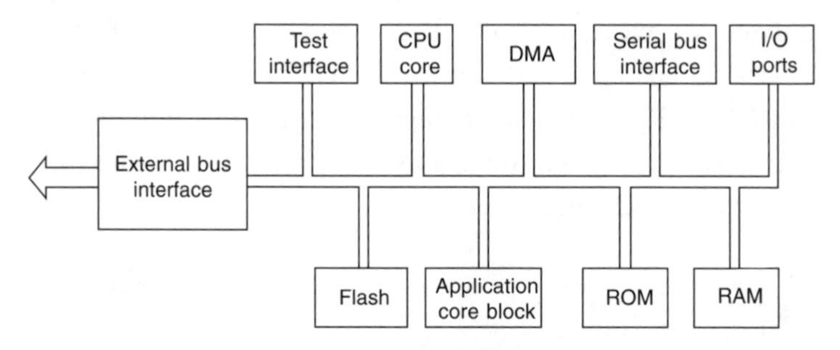

Figure 2.22
System-on-a-chip

In short, a SoC doubles up as a general purpose and a dedicated processor.

Although SoC is the best-known method of combining processor core with system applications, it requires a high production volume to make it viable. A new generation of system-level integration, known as *system-in-package* (*SiP*), has been developed, which is capable of higher integration capacity. It reduces the number of components in the system and simplifies the design of the main board. The reduction or even elimination of many passive components, such as resistive transistors and local bypass capacitors, is another distinctive design benefit.

Analogue-to-digital conversion

In most applications the initial information, such as video or audio, is in the form of an analogue signal, which must be converted into a digital format before digital processing can take place.

Analogue-to-digital conversion consists of two distinct stages: sampling and quantizing (Figure 2.23). *Sampling* is the process of assessing the value of the analogue signal at regular intervals. The samples are then rounded off to the nearest predetermined level, a process known as *quantization*. The amplitude of each sample may be represented by a binary code, a process known as pulse code modulation (*PCM*). The number of bits used to identify each pulse is determined by the number of quantized levels – for instance, 8 different levels may be identified using 3 bits, while 64 levels require 6 bits, and so on. Conversely, the number of available quantized levels is determined by the bit width of the converter – a 3-bit converter, for instance, provides $3^2 = 8$ levels, and an 8-bit converter provides $8^2 = 256$ levels, and so on. Table 2.9 lists all the possible outputs of a 3-bit ADC with a quantum of 0.25 V. At the receiving end, the samples are reproduced at their quantified levels and their peaks are joined to reconstruct the original analogue signal (Figure 2.24).

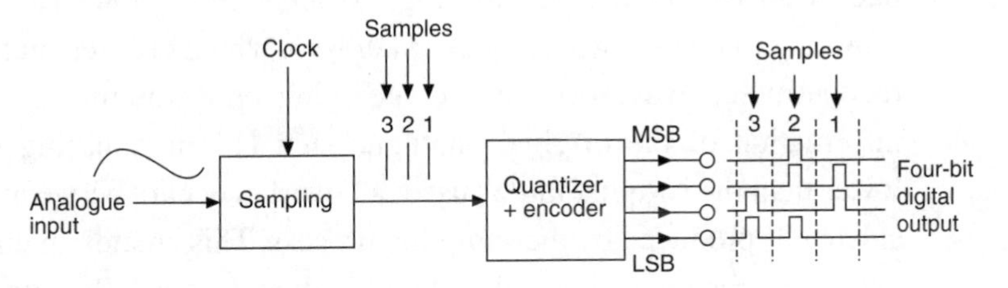

Figure 2.23
Analogue-to-digital conversion

Sampling rates

The Shannon sampling theory states that sampling an analogue signal does not remove any information provided that the sampling

Table 2.9 Quantized levels

Level	*Sample voltage (V)*	*Binary code*		
		MSB		*LSB*
0	0	0	0	0
1	0.25	0	0	1
2	0.50	0	1	0
3	0.75	0	1	1
4	1.00	1	0	0
5	1.25	1	0	1
6	1.50	1	1	0
7	1.75	1	1	1

Figure 2.24
Waveform reconstruction

rate is no less than twice the highest analogue frequency. This sampling rate is known as the *Nyquist rate*, which ensures that the reconstructed waveform at the receiving end contains all the information of the original analogue signal. If a sampling rate lower than the Nyquist rate is used, an overlap occurs between the sidebands produced by the sampling process. This creates an effect known as ailising, which makes it impossible to recover the original signal without distortion. However, recovering signals that have been sampled at the Nyquist rate requires an ideal filter with very sharp cut-off characteristics, and this situation must be avoided. In practice, a sampling rate that is slightly (about 10 per cent) higher than the Nyquist rate is used to provide enough separation between the sidebands, allowing the employment of practical filters.

Quantizing error

With the input being analogue, sample values will invariably fall between the predetermined discrete quantized levels and hence there is always an element of uncertainty in terms of the actual value of the least significant bit (LSB). This ambiguity gives rise to what is known as the *quantizing error*, an error that is inherent in any digital coding of analogue values. The effect of this type of noise is that some data bits are received incorrectly, thus distorting the reconstructed waveform.

Quantized error has a constant quantity equal to one-half the quantum, and the effect of quantized noise is therefore more noticeable when the analogue signal is at low level resulting in poor signal-to-noise (S/N) ratio. For instance, assuming a quantum level of 250 mV, the quantized error will be

$$0.5 \text{ quantum} = 0.5 \times 250 = 125 \text{ mV} = 0.125 \text{ V}$$

If the analogue signal is strong, say 1.5 V, then

$$\text{S/N ratio} = 1.5/0.125 = 12, \text{ i.e. } 20 \log 12 = 21.6 \text{ dB}$$

For a weak analogue signal with an amplitude of, say, 0.25 V, then

$$\text{S/N ratio} = 0.25/0.125 = 2, \text{ i.e. } 20 \log 2 = 6 \text{ dB}$$

Bit rate

The bit rate is defined as the number of bits produced per second by the encoder. Given a sampling frequency of f_s and an n-bit quantizer, then the bit rate generated is given by

$$\text{Bit rate} = n \times f_s \text{ bits per second}$$

For example, an 8-bit ADC with a sampling frequency of 10 MHz generates a bit rate of 8 × 10 MHz = 80 Mbits/s.

Bandwidth requirements

The bandwidth requirement of a communication channel is the maximum frequency that may be produced by the bitstream. That maximum frequency is determined by the bit rate generated by the quantizing encoder. The maximum rate of change of a bitstream (i.e. the maximum frequency) is obtained when adjacent bits alternate between 0 and 1, as shown in Figure 2.25. One complete cycle of the waveform with a periodic time T contains two bits, and it follows therefore that the frequency of the waveform is half the bit rate, giving a bandwidth = 0.5 bit rate. In other words, a channel with a bandwidth of B has a capacity of 2B pulses per second. This is the theoretical value; in practice, this capacity is reduced somewhat to kB where k is between 1 and 2 depending on, among other things, the shape and width of the pulses.

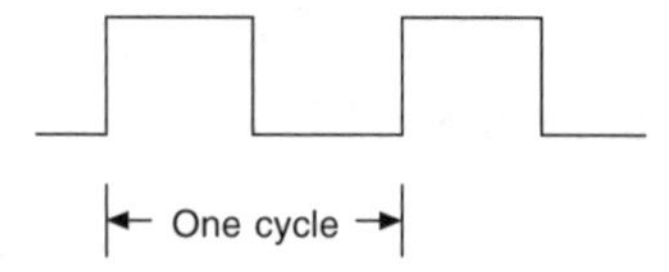

Figure 2.25

Channel capacity

Up to 1948, when Shannon's *Mathematics Theory of Communications* was published, it was assumed that, because of the presence of random noise in a channel, error-free transmission could only be achieved if the bit rate approaches zero. However, Shannon proved that random noise does not by itself set any limit on the accuracy of the transmission. Instead, it sets a limit on the rate of information (the bit rate) that can be achieved in a channel in which the probability of error is very small or approaching zero – what is known as a *virtual error-free communication* system. Shannon proved that the limitation imposed on the capacity (C) of a channel is determined by the available bandwidth (B) and the signal-to-noise ratio (S/N):

$$C = B \log_2 (1 + S/N) \text{ bits per second}$$

If there were no noise on the channel (N = 0), there would never be any uncertainty in the received pulses and thus capacity (C) would be infinite. Communications would then cease to be a problem. However, for a given signal-to-noise ratio, the bandwidth determines the bit rate that can be achieved for an error-free environment. Conversely, given a channel bandwidth, the signal-to-noise ratio determines the maximum bit rate that may be achieved. It follows that bandwidth and signal-to-noise ratio are interchangeable and may be traded against each other in the design of any system of communication. The relationship between bandwidth and signal-to-noise ratio is logarithmic, whereby a relatively small increase in bandwidth may be traded for a large improvement in signal-to-noise ratio and consequently a large reduction in transmitted signal power.

Non-linear quantization

Although the quantization error cannot be wholly avoided, it can be minimized by improving the resolution of the converter through increasing its bit width – thus reducing the quantized level, the quantization error and the quantization noise. However, this still leaves weak signals with a comparatively poor signal-to-noise ratio. To overcome this *non-linear quantizing* may be used, in which the quantum level for weak signals is decreased (i.e. more bits are utilized to represent them) compared with the quantum level for strong signals. This type of non-linear quantization, known as *companding*, tends to equalize the signal-to-noise ratio over the range of sample amplitudes generated by the analogue signal. At the receiving end, a complementary non-linear DAC is employed to reproduce the original analogue signal. Another technique, which achieves a similar result, is to use an analogue voltage compressor to precede the linear encoder at the transmitting end – a process known as *pre-emphasis*. The compressed analogue voltage gives prominence to weaker signals. At the receiving end the process is reversed, using a voltage expander.

In many applications, such as digital audio or video processing, linear quantization produces a bit rate that may be higher than can be accommodated by the available bandwidth. To rectify this a controlled quantizer is used, which can vary the resolution of the encoder to ensure a constant bit rate at its output. This is carried out dynamically, as the input data stream varies in quantity and speed.

Error control techniques

In all types of communication systems, errors may be minimized but they cannot be avoided completely – hence the need for error detection and/or correction techniques. If an error is detected at the receiving end, it can be corrected in two different ways: the recipient can request the original transmitter for a repeat of the transmission (known as feedback or backward error correction), or the recipient can attempt to correct the errors without any further information from the transmitter (known as forward error correction). Retransmission is commonplace in communications systems where it is possible. However, if distances are large (such as communications with a space probe) or real time signals are involved (such as in audio and video broadcasting), retransmission is not an option. In these cases, error correction techniques must be employed.

Information messages invariably contain what is known as *redundancy.* In ordinary English text, the U following the Q is quite unnecessary and 'At this moment in time' can be easily reduced to 'At this moment' or even 'Now'. Redundant letters or words play a very important role in communication, in that they allow the recipient to make sense of distorted information. This is how it is possible to make sense of badly spelled seaside postcards, a corrupted fax message, a badly tuned radio or television, and so on.

As far as digital communication is concerned, redundancy is unnecessary data that occupy precious bandwidth space. For this reason, compression techniques are used to ensure that only necessary data are transmitted. In video and audio processing, raw data are compressed by 100 times or more to reduce the bandwidth requirements. To provide for error correction capability, controlled redundancy (i.e. extra bits) is added to enable messages corrupted in transmission to be corrected at the receiving end.

The most basic error control technique is parity. *Parity* involves a single parity bit at the end of a digital word to indicate whether the number of ones is even or odd (Figure 2.26). There are two types of parity checking: even parity (Figure 2.26a) is when the complete coded data set (including the parity bit) contains an even number of ones, whereas odd parity (Figure 2.26b) is when the complete coded data wet contains an odd number of ones. At the receiving end, the number of ones is counted and checked against the parity bit; a difference indicates an error. This simple parity check can only detect an error occurring in a single bit – an error affecting two bits will go undetected. Furthermore, there is no provision for determining which bit is actually faulty. For this reason, more

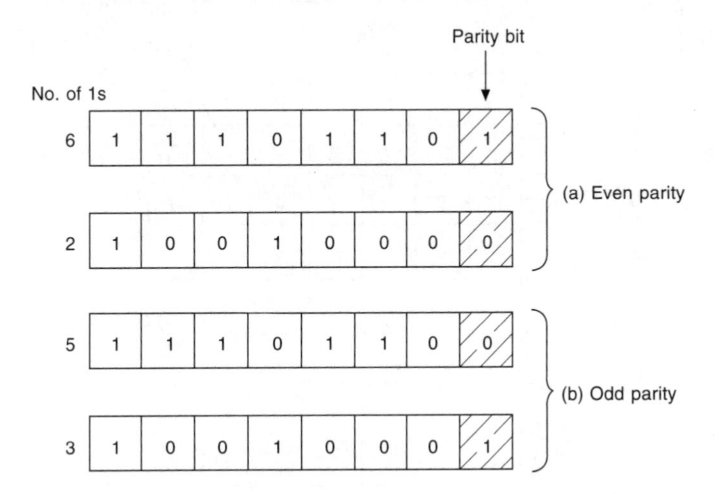

Figure 2.26
Parity

sophisticated techniques are normally used with high levels of redundancy.

Block coding

In the simplest form of *block coding*, serial data are first rearranged into blocks of rows and columns (Figure 2.27a). Parity bits for instance even parity are then added to each row and column as

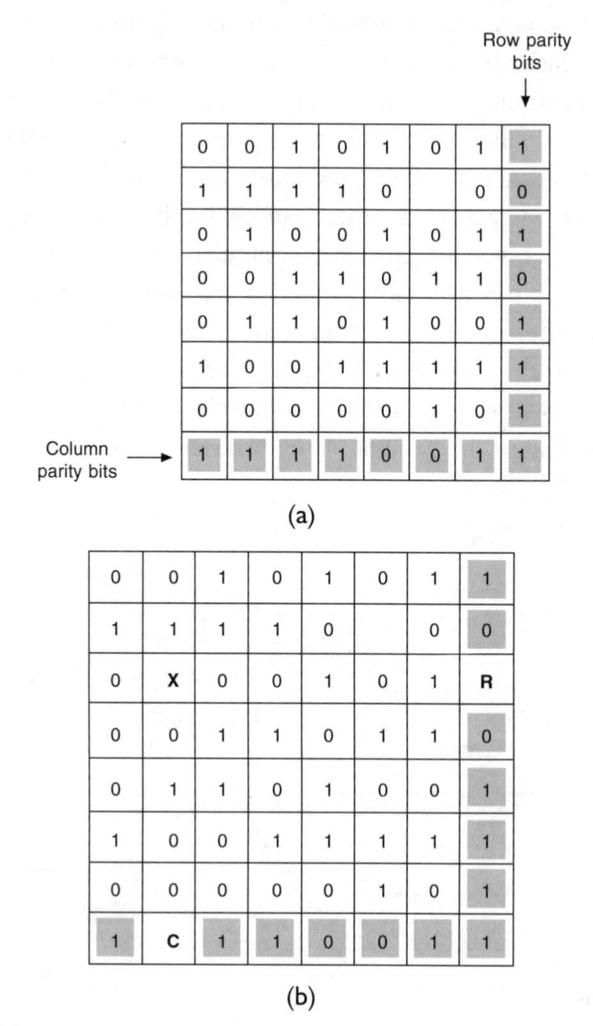

Figure 2.27

shown. A single bit error will cause two parity bits to indicate an error. In Figure 2.27b, bits R and C indicate an error in the relevant row and column, which points to an erroneous bit X.

Two main factors determine the type of error correction scheme used: the *bit error rate* (*BER*) and the type of errors expected. The latter refers to whether errors are expected to occur in single bits or in bursts; burst errors may be overcome by bit and block interleaving. The BER is the ratio of the number of bits that are likely to be corrupted in a specified bitstream. For instance, a BER of 10^{-10} means that 1 bit in 10^{10} bits may be corrupted.

In DVD applications a more sophisticated and robust system of error correction is employed, the details of which are beyond the scope of this book.

Interleaving

In many communication systems, errors occur in bursts. Correcting such errors is more difficult than correcting single bit errors, and to overcome this difficulty *interleaving* is used. Two types of interleaving are used in digital television applications; bit interleaving, and block or symbol interleaving.

Bit interleaving is the process of rearranging the sequence of bits before transmission takes place. The principles of bit interleaving are illustrated in Figure 2.28, in which the order of an 8-bit word

Original bit sequence	0	1	2	3	4	5	6	7
After interleaving	2	5	7	1	4	6	0	3
Wrong bits due to errors				X	X	X		
Damaged bit sequence	2	5	7	X	X	X	0	3
Restored original bit sequence after deinterleaving	0	X	2	3	X	5	X	7

Figure 2.28
Bit interleaving

(b_0–b_7) is rearranged by interleaving to b_2, b_5, b_7, b_1, b_4, b_6, b_0 and b_3. If after transmission three adjacent bits (b_1, b_4, b_6) were faulty, then de-interleaving at the receiving end would restore the original order of the bits and separate the error burst into single bit errors as shown – which may then be corrected by such coding techniques as Hamming or Reed-Solomon.

In *block interleaving* (illustrated in Figure 2.29), codewords produced by the encoder are written in a memory buffer, row by row – C_1, C_2, C_3, and so on. When the rows have been filled, the codewords are extracted column by column – C_{13}, C_9, C_5, C_1, C_{14}, and so on.

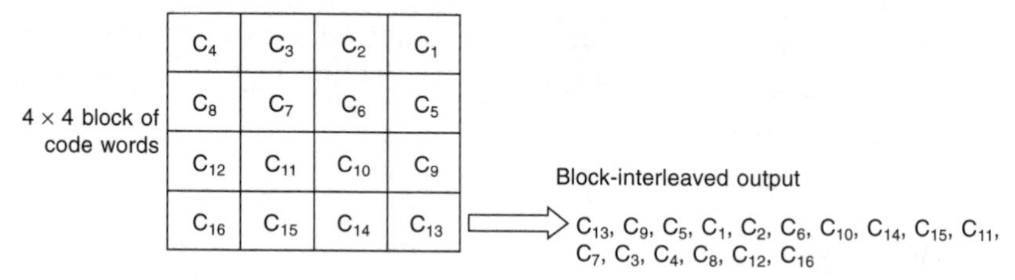

Figure 2.29
Block interleaving

Scrambling and encryption

Scrambling is the rearrangement or transposition of the order of the data bits in a predetermined manner, and is not to be confused with encryption, which is the replacement of the original information by an alternative code pattern. *Encryption*, being a secure system, is used in conditional access applications where program restriction applies, while scrambling is a system used only for energy dispersal. In general, however, scrambling is used for both processes, and the two terms have become interchangeable.

The problem with plain unscrambled data streams is that they are likely to have long series of zeros and ones, which introduce a DC component, resulting in an uneven distribution of energy making the transmission highly inefficient. If the bitstream can be randomized and the series of zeros and ones scattered, a more even energy distribution will be obtained. This is the purpose of scrambling. Total random scattering is not possible, as there is no way of descrambling the bits back to their original order at the receiver. However, a *pseudo-random* scattering with a known and predictable pattern can be easily descrambled at the receiving end to regenerate the original order of the data bits. This is achieved using a pseudo-random-bit sequencer, which produces the same result as total random distribution. The scrambler consists of a feedback shift register whose output is predictable. At the receiving end, a reciprocal process takes place using a forward shift register, which descrambles the data bits back to their original order. In order to ensure that the start of each transport packet can be recognized, the start byte is not scrambled and the scrambler is disabled for the duration of the start code.

Oscillators and phase-locked loop

One of the requirements of a digital and microprocessor system is an accurate clock pulse, and such a pulse may be obtained from a fixed-frequency crystal-controlled oscillator. The oscillator is basically an amplifier with a large enough positive feedback for an output to be produced without an input (Figure 2.30). If a quartz crystal is placed in the feedback loop, the frequency of the output

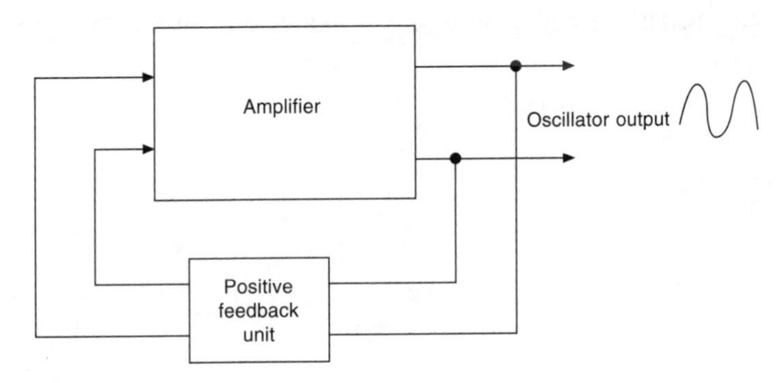

Figure 2.30

will be determined by the highly precise natural frequency of the crystal. However, in most video and audio operations a clock is required that is not only accurate, but also run synchronously with an external source. To achieve this, a *phase-locked loop* (*PLL*) is employed. In the PLL the voltage-controlled oscillator (VCO) is used, as illustrated in Figure 2.31.

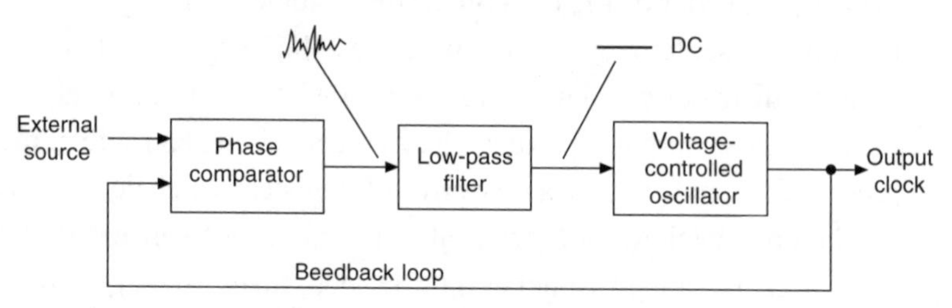

Figure 2.31

The frequency of a voltage-controlled oscillator is determined by the DC voltage applied to it – hence the name. The DC voltage is derived from a phase comparator followed by a low-pass filter. The phase comparator is fed with two signals: an input signal representing the frequency of the external source, and a feedback signal representing the output. The phase comparator produces a voltage that represents the phase difference of the two signals. This voltage is then fed into a low-pass filter to remove any random or transient changes and the resulting DC is fed into the VCO to set the frequency of the output clock signal. The loop is completed with the output clock signal fed back to the phase comparator as shown. Any change in the external source frequency is then reflected

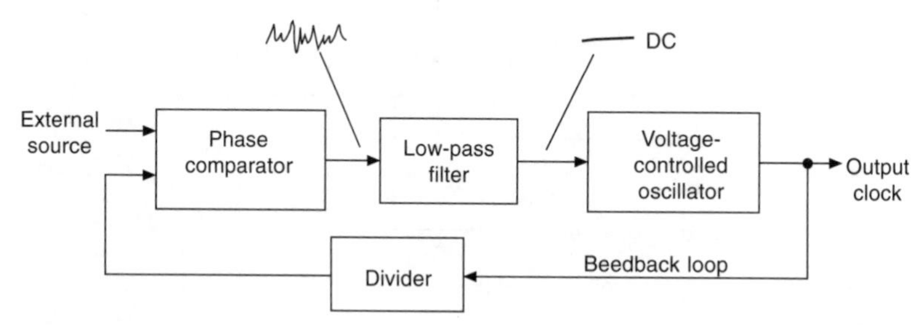

Figure 2.32

in a change in the DC output from the phase comparator, causing a compensating change in the frequency of the voltage-controlled oscillator to keep the output clock in sync with the external source. Normally, a divider is included in the feedback loop as shown in Figure 2.32 to provide a facility for changing the frequency of the output clock by merely changing the divide-by factor of the divider.

CHAPTER 3

VIDEO SIGNALS: ANALOGUE AND DIGITAL

Introduction

A video signal consists of a waveform that explores a scene or a picture along a number of horizontal lines known as *scan lines*. For a moving picture, the process of scanning a scene has to be repeated several times a second. These waveforms are produced by television cameras. At a television studio, the scene is projected on a photosensitive plate located inside the TV cameras and is repeatedly scanned by a very fast electron beam, which ensures that consecutive images differ only slightly. At the playback end, a *cathode ray tube* (*CRT*) is used to recreate the picture by an identical process of scanning a coated screen with a moving electron beam. The phenomenon of persistence of vision then gives the impression of a moving picture in the same way as a cine film does.

Scanning

In order to explore the scene in detail, the brightness of each 'element' of the picture frame is examined line by line (see Figure 3.1). The electron beam sweeps across the scene from left to right, and then returns very quickly (the '*flyback*') to begin the next scan

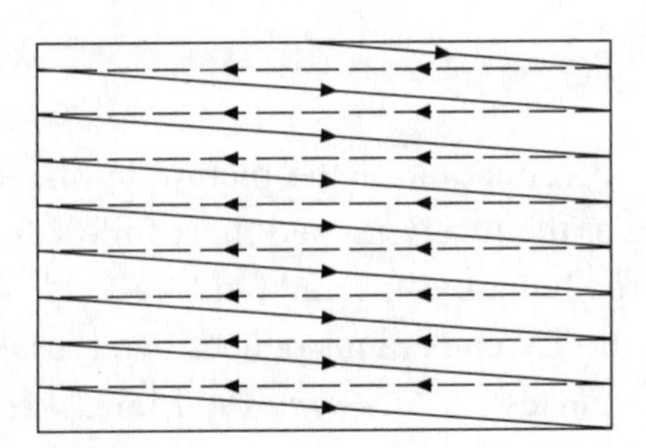

Figure 3.1
Line scanning

line and so on. The number of lines scanned varies depending on the required picture definition. Standard TV broadcasting in the UK employs a total of 625 lines, while in the USA 525 lines are used. High definition television (HDTV) uses 1250 lines or more. At the end of each picture scan the electron beam moves back to the top of the scene and the sequence is repeated. With 625 lines per picture and 25 pictures per second, the scan-line frequency may be calculated as $25 \times 625 = 15\,625$ Hz, or 15.625 kHz. In the USA a rate of 30 pictures per second is used, giving a line frequency of $30 \times 525 = 15.75$ kHz.

There are two types of scanning, sequential and interlaced. *Sequential scanning* involves scanning complete pictures, also called frames, at one time (625 lines), followed by another complete picture scan. *Interlaced scanning*, on the other hand, involves scanning the 'odd' lines (1, 3, 5, etc.) followed by the 'even' lines (2, 4, 6 etc.) – and thus only half of the picture, known as a *field*, is scanned each time. A complete picture therefore consists of two fields, known as top (or odd) and bottom (or even), resulting in a field frequency of $2 \times 25 = 50$ Hz. Interlaced scanning avoids the flicker on the television receiver that is associated with the 25-Hz sequential scanning.

In the absence of picture information, scanning produces what is known as a *raster*.

Sync pulses

For faithful reproduction of the picture by the cathode ray tube, the scanning at the receiving end must follow the scanning at the transmitting end line-by-line and field-by-field. To ensure that this takes place, the TV camera introduces two types of synchronizing pulses (sync pulses): a *line sync* (at 15.635 kHz for the UK) to indicate the start of a new scan line, and a *field sync* (at 50 Hz for the UK) to indicate the start of a new field.

Composite video waveform

When the sync pulses are added to the picture information, a *composite video* waveform is produced (Figure 3.2). The picture information is represented by the waveform between the line sync pulses, and thus may acquire any shape depending on the picture brightness along the scan line. The waveform shown represents a line that starts at peak white (maximum brightness) and decreases in brightness to black level (minimum brightness) in eight steps. Such a waveform represents what is known as *grey scales*.

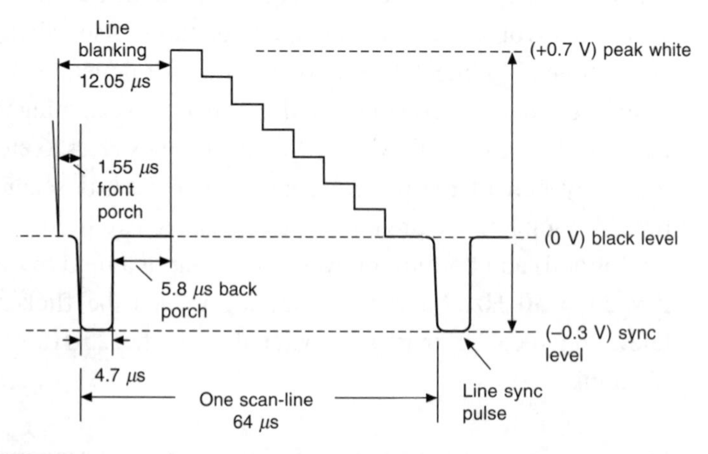

Figure 3.2
Composite video for one scan-line

The total available voltage is divided into two regions:

1. Below black level region, 0–0.3 V, reserved for sync pulses (line and field)
2. Above black level region, 0–0.7 V (peak white), used for video or picture information.

Before and after every sync pulse the voltage is held at black level for short periods of time, known respectively as the front porch and the back porch. The *front porch* has duration of 1.55 μs; it ensures the video information is brought down to the black level before the sync pulses is applied. The *back porch* has a longer duration of 5.8 μs; to provide time for the flyback to take place before the application of the video information. As can be seen, the front porch, the sync pulse and the back porch are at or below the black level. During this time, a total of 12.05 μs, the video information is completely suppressed; this is known as the line blanking period. For this reason, composite video is frequently referred to as *CVBS* (composite video, blanking and sync).

The duration of one complete line of composite video may be calculated from the line frequency:

Line duration = 1/line frequency = 1/15.625 kHz = 64 μs

Pixels

The frequency of the video waveform is determined by the change in the brightness of the line as the electron beam scans it. Maximum video frequency is obtained when adjacent active elements, known as *pixels*, are alternately black and white (Figure 3.3); this represents the maximum definition of a TV image. The number of pixels along a vertical line is determined by the number of scan lines, which is nominally 625 for the UK. However, a number of lines are rendered inactive owing to blanking of the picture during the flyback period. This reduces the number of *active lines* to 576.

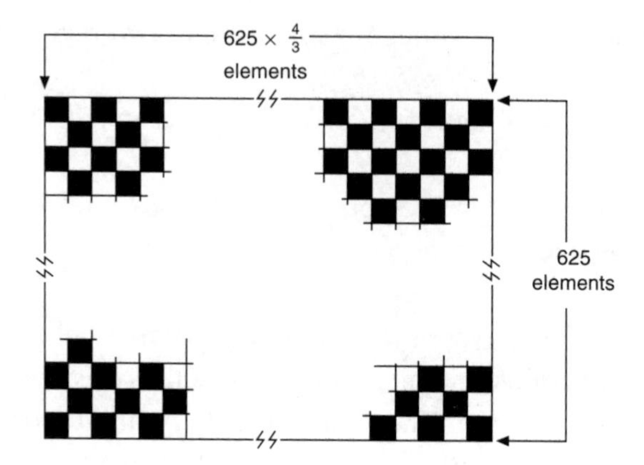

Figure 3.3
Maximum video frequency is obtained when adjacent pixels are alternately black and white

For equal definition along a horizontal line, the separation between the black and white pixels must be the same as the separation along the vertical line. For a perfectly square television screen, an equal number of pixels would be required in both directions. However, the TV screen has an aspect ratio of 4 : 3. The number of horizontal pixels must thus be increased to $576 \times (4/3) = 768$, giving a picture total of

$$576 \times 768 \times 4/3 = 442\ 368 \text{ pixels}$$

Video bandwidth of analogue video

When an electron beam scans a line containing alternate black and white pixels, the video waveform is as shown in Figure 3.4 – representing the variation in brightness along the scan line. It can be seen that for any adjacent pair of black and white pixels, one complete cycle is obtained. Hence for the 10 pixels shown, five complete cycles are produced. It follows that for a complete picture of alternate black and white pixels, the number of cycles produced is

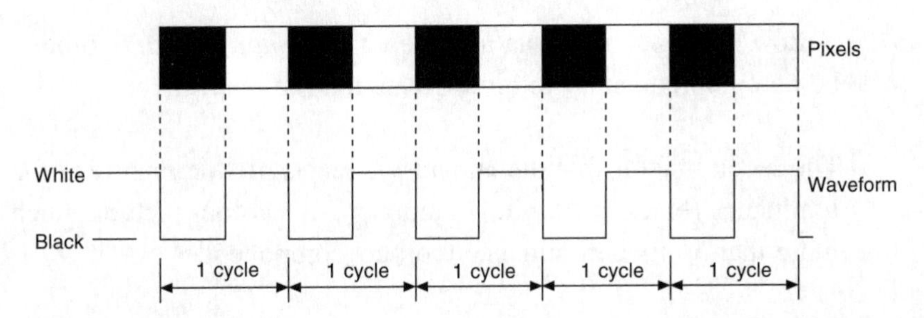

Figure 3.4
Video waveform for alternate black and white elements

$$0.5 \times \text{total number of pixels} = 0.5 \times 442\,368$$
$$= 221\,184 \text{ cycles per picture}$$

Since there are 25 complete pictures per second,

$$\text{Number of cycles per picture} = 221\,184 \times 25$$
$$= 5\,529\,600 \text{ Hz} = 5.53 \text{ MHz}$$

The minimum video frequency is obtained when the electron beam scans pixels of unchanging brightness. This corresponds to unchanging amplitude of the video waveform – a frequency of 0 Hz or DC. The overall bandwidth is therefore 0 Hz to 5.53 MHz. In practice, 5.5 MHz is normally used.

Principles of colour television

The principle of colour transmission is based on two facts. The first is that all colours may be produced by the addition of appropriate quantities of the three *primary colours* – red, green and blue (RGB). For example:

Yellow = R + G

Magenta = R + B

Cyan = B + G

White = R + G + B

Yellow, magenta and cyan are known as *complementary* colours, being complementary to blue, green and red respectively.

The second fact is that the human eye reacts predominantly to the luminance (black and white) component of a colour picture, much more than to its chrominance (colour) component.

Colour TV transmission involves the simultaneous transmission of the luminance and the chrominance components of a colour picture with the luminance part predominant over the chrominance. In analogue TV broadcasting, the *luminance signal Y* (representing the luminance component) is transmitted directly in the same way as a monochrome transmission. This provides the necessary compatibility between the two systems. As for the chrominance component, it is first 'purified' by removing the luminance component from each primary colour, resulting in what is known as *colour difference* signals:

R – Y

G – Y

B – Y

Since the luminance signal Y = R + G + B, only two colour difference signals need to be transmitted, namely R – Y and B – Y. The third colour difference (G – Y) may be recovered at the receiving from the three transmitted components: Y, R – Y and B – Y. In analogue TV broadcasting, the two colour difference signals R – Y and B – Y are known as *U* and *V* respectively. In digital applications, they are referred to as C_R and C_B.

The PAL colour system

There are three main systems of analogue colour encoding NTSC, PAL and SECAM. All three systems split the colour picture into luminance and chrominance, and all three systems use colour difference signals to transmit the chrominance information. The difference between the three systems lies in the way in which the subcarrier is modulated by the colour difference signals. SECAM (used in France) transmits the colour difference signals on alternate lines. The other two systems, NTSC (used in the USA) and PAL (used in the UK), employ both chrominance components simultaneously using quadrature amplitude modulation. However, errors in hue may occur as a result of phase errors (delay or advance) of the chrominance phasor. Such errors are caused either by the receiver itself or by the way in which the signal is propagated. They are almost completely corrected by PAL.

In the *PAL* (phase alternate line) system, the V signal is reversed on successive lines so that V on one line is followed by –V on the next and so on. The first line is called the NTSC line and the second is called a PAL line. Phasor errors are thus reversed from one line to the next. At the receiving end, a process of averaging takes place either by the human eye (*PAL–S*) or by employing a delay line to allow for consecutive lines (NTSC and PAL) to be added to each other, thus cancelling out any phase errors. This latter technique is known as *PAL–D*. The effect of sending NTSC video signals to a PAL television receiver is the loss of colour.

Digitizing the TV picture

Digital video applications involve processing moving pictures as well as high quality six-channel sound. Processing such high a volume of information in the form of digital bits results in a very high-speed data stream, and to avoid this, data compression is used. The first step is to transform the analogue video signal into a digital format.

Digitizing a TV picture is the process of sampling the contents of a picture frame, scan line by scan line (Figure 3.5). In order to maintain the quality of the picture there must be, at a minimum, as

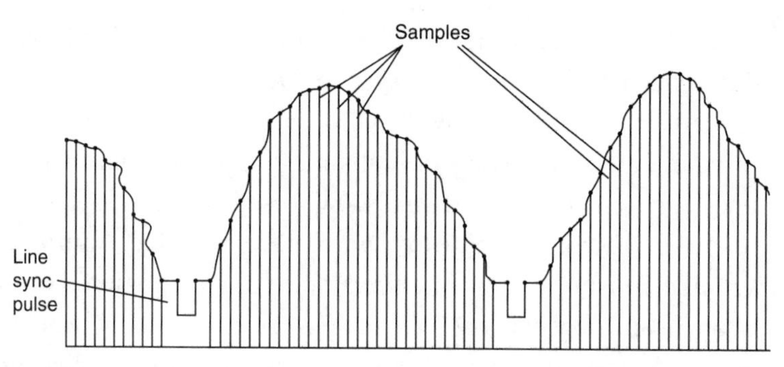

Figure 3.5
Video sampling

many samples per line as there are pixels, with each sample representing 1 pixel. There are two factors that determine the number of pixels in a television picture: the number of lines per picture, and the number of pixels per line. The British PAL system uses 625 lines, of which 576 are 'active' in that they may be used to carry video information. (For the American 525-line system, the corresponding figure is 480.) The number of pixels allocated per line is 720, giving a total number of pixels per picture of $576 \times 720 = 414\ 720$.

Each line will therefore be represented by 720 samples, with each sample representing 1 pixel – sample 1 representing pixel 1; sample 2 representing pixel 2, etc. (Figure 3.6). The process is then repeated for the second line and so on to the end of the frame, when it is

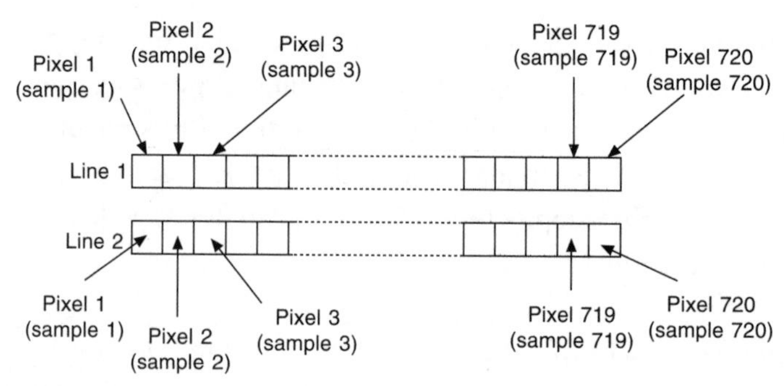

Figure 3.6
Line pixels

repeated all over again for the next frame. To ensure that the samples are taken at exactly the same physical point from frame to frame (Figure 3.7), the sampling frequency must be locked to the line frequency. For this to happen, the sampling rate must be an exact multiple of the line frequency.

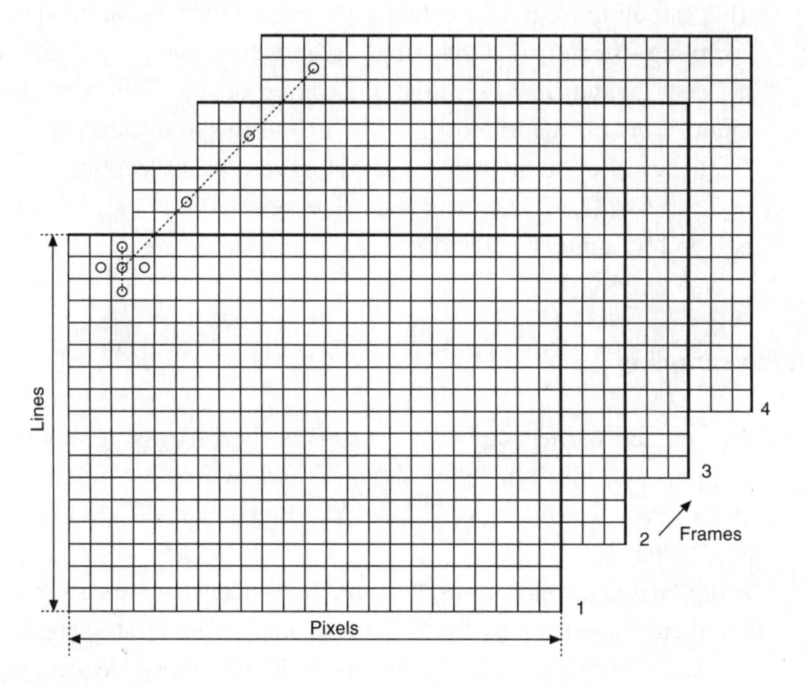

Figure 3.7
Samples must be taken at the same point in each frame

The sampling rate

Of the total 64-μs period of one line of composite video, 12 μs is used for the sync pulse, the front porch and the back porch, leaving 52 μs to carry the video information. With 720 pixels per line,

$$\text{Sampling rate} = \text{number of pixels per line}/52\ \mu\text{s}$$
$$= 720/52 = 13.8\ \text{MHz}$$

However, since the sampling frequency must be a whole multiple of the line frequency, a sampling rate of 13.5 MHz (864 and 858 $\times$ line frequency of the 625 and 525 broadcasting systems respectively) is recommended by the CCIR (Comité Consultatif International Radiocommunication).

This sampling frequency satisfies the other criterion for adequate sampling – namely that the sampling frequency must be at least 10 per cent greater than twice the highest frequency of the analogue input. Thus for the video signal, which at the studio may have a frequency of up to 5.5 MHz, a sampling rate 10 per cent in excess of $2 \times 5.5 = 11$ MHz is necessary. The selected rate of 13.5 MHz is therefore quite satisfactory.

Video sampling

As outlined earlier, colour TV involves the processing of three components; the luminance signal Y and two colour difference signals, Y – R (known as C_R) and Y – B (known as C_B).

In digital video applications, the three components are first sampled and then converted into digital data streams before processing. For the luminance signal, which contains the highest video frequencies, the full sampling rate of 13.5 MHz is used. As for chrominance components C_R and C_B, which contain lower video frequencies and to which the human eye is less sensitive, a lower sampling rate is acceptable. The CCIR recommends a sub-sampling rate for the chrominance of half the luminance rate, i.e. $0.5 \times 13.5 = 6.75$ MHz, which gives a total sampling rate of $13.5 + 6.75 + 6.75 = 27$ MHz. This frequency is known as the *system clock* frequency.

The 4 : 2 : 0 sampling structure

There are several ways of sub-sampling the chrominance components. One technique provides for the chrominance components to be sampled every other pixel. This technique reduces the chrominance resolution in the horizontal dimension only, leaving the vertical resolution unaffected. It is known as the 4 : 2 : 2 sampling structure, with the ratio 4 : 2 : 2 indicating that both C_R and C_B are sampled at half the rate of the luminance signal.

For smaller bandwidth requirements (and hence a lower bit rate), a 4 : 1 : 1 sampling structure may be used. In this sampling technique the chrominance components are sampled at a quarter of the luminance rate – hence the ratio 4 : 1 : 1. The 4 : 1 : 1 sampling framework was used in early digital applications and produced good results; however, it does suffer from a large imbalance between the vertical and horizontal chrominance resolution. To overcome this while keeping the same bit rate, MPEG decided to reduce the chrominance resolution by equal amounts in both the horizontal and vertical dimensions, and that is how the 4 : 2 : 0 sampling came about.

The bit rate

Sampling is followed by the quantization block, where the sample values are rounded up or down to quantum values before they are converted into a multi-bit code. For studio applications, 10-bit coding is used; for domestic applications such as TV broadcasting and DVD, 8-bit coding is regarded as adequate for good quality picture reproduction. An 8-bit code provides $2^8 = 256$ discrete signal levels.

The bit rate may be calculated as follows:

Bit rate = number of samples per second × number of bits/sample

but

Number of samples/s = number of samples per picture times number of pictures/s, and

$$\text{Number of samples per picture} = 720 \times 576 = 414\ 720$$

Given a picture rate of 25,

$$\text{Number of samples/s} = 720 \times 576 \times 25 = 1\ 036\ 800$$

therefore

Bit rate generated by the luminance component using an 8-bit code $= 720 \times 576 \times 25 \times 8 = 82\ 944\ 000 = 82.944$ Mbits/s

The bit rate for the chrominance components depends on the chroma sub-sampling rate as determined by the sampling structure used. For a 4 : 2 : 2 sampling framework with horizontal sub-sampling only,

$$\text{Number of samples per picture} = 360 \times 576 = 207\ 360$$

which gives a chrominance bit rate of

$$\begin{aligned}
&360 \times 576 \times \text{picture rate} \times \text{number of bits} \times \text{number of} \\
&\text{chrominance components} = 360 \times 576 \times 25 \times 8 \times 2 \\
&\qquad = 82\ 944\ 000 \\
&\qquad = 82.944\ \text{Mbits/s}
\end{aligned}$$

Therefore, the total bit rate for 4 : 2 : 2 sampling is

$$\begin{aligned}
\text{Luminance bit rate} + \text{chrominance bit rate} &= 82.944 + 82.944 \\
&= 166\ \text{Mbits/s}
\end{aligned}$$

For a 4 : 2 : 0 sampling framework where both horizontal and vertical sub-sampling is used, the chrominance bit rate is

$$360 \times 288 \times 25 \times 8 \times 2 = 41\ 472\ 000 = 41.472\ \text{Mbits/s}$$

thus

Total bit rate = 82.944 + 41.472 = 124 Mbits/s

Considering that the bit rate provided by the fastest modem is 56 Kbits/s and that even the modern SDL lines provide no more than 10 Mbits/s, 124 Mbits/s is an impractically high bit rate. This is the reason for data compression.

Picture quality

MPEG-2 defines a number of digital video quality standards by specifying a set of profiles and levels. A *profile* defines the degree of complexity of the encoding process and the decoding system, while a *level* describes the picture properties such as picture size, resolution and bit rate. MPEG-2 defines six profiles and four levels. Of the 24 possible combinations, only 13 are defined by MPEG (Figure 3.8). For entertainment purposes, what is known as standard digital television (SDTV) Main Profile at Main Level (*MP@ML*) is used. This specifies the number of pixels as 720 × 576 and the sampling structure as 4 : 2 : 0.

Video formats

Early in the development of motion pictures, the film industry agreed on a standard frame proportion of 4 : 3 or 1.33 : 1. The same aspect ratio was adopted for television screens in the late 1940s, but the advent of cinemascope saw the introduction of widescreen movies with aspect ratios of 1.85 : 1 or 2.35 : 1. The television industry retained the traditional 4 : 3 format until relatively recently, when widescreen television receivers with an aspect ratio of 16 : 9 (or 1.78 : 1) were introduced. The widescreen aspect ratio was then adopted by digital television and subsequently by DVD-video. The new widescreen format retains the 720 × 576-pixel format (720 × 480 for NTSC), and with it the bit rate requirements of the traditional 4 : 3 format (Figure 3.9).

Level \ Profile	Simple	Main	SNR Scalable (Note 2)	Spatial scalable	High
High 1920 × 1080 × 30 or 1920 × 1152 × 25 (Note 1)		MP@HL US digital HDTV			HP@HL
High - 1440 1440 × 1080 × 30 or 1440 × 1152 × 25 (Note 1)		MP@H1440		SSP@H1440 European digital HDTV	HP@H1440
Main 720 × 480 × 29.97 or 720 × 576 × 25 (Note 1)	SP@ML Digital transmission cable TV	MP@ML (Note 3) DVD-Video, Digital satellite broadcasting (PerfecTV and others)	SNP@MP		HP@ML
Low 352 × 288 × 29.97 (Note 1)		MP@LL	SNP@LL		

Figure 3.8
Currently prescribed profiles/levels

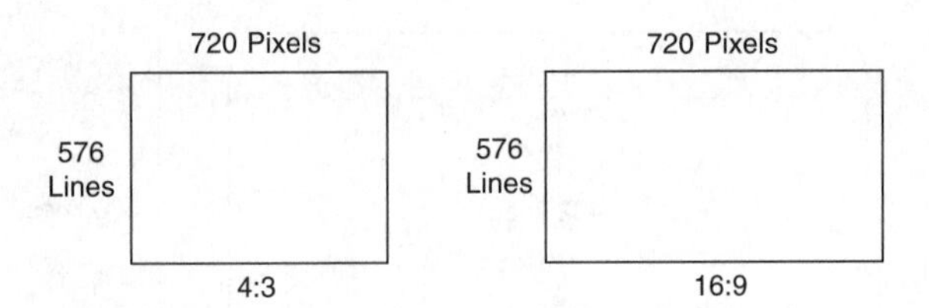

Figure 3.9

The 16 : 9 format was chosen in part because it is an exact multiple of 4.3 (4/3 × 4/3 = 16/9). This simple relationship makes conversion from one format to the other relatively easy. Furthermore, a 16 : 9 (1.78 : 1) format is very close to the format used by most films, namely 1.85 : 1, which again makes conversion fairly straightforward.

At DVD playback the shape of a displayed image depends on two factors – the original format, and the format selected by the user. There are four different ways in which a picture may be displayed:

1. Full frame, where the material is shot in 4 : 3 and displayed in full 4 : 3 format
2. Pan & scan (Figure 3.10), where the picture, regardless of its shape, is made to fill the 4 : 3 format by selecting and displaying the most important action in the frame (known as centres of interest), leaving out material on either side
3. Letterbox (Figure 3.11), which is a method of showing widescreen on a 4 : 3 format screen. The widescreen (16 : 9 or wider) is made to fit the screen by adding mattes to the top and

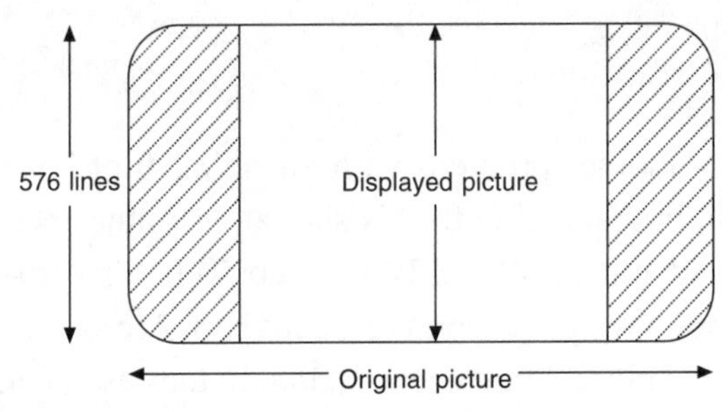

Figure 3.10
Pan & scan display

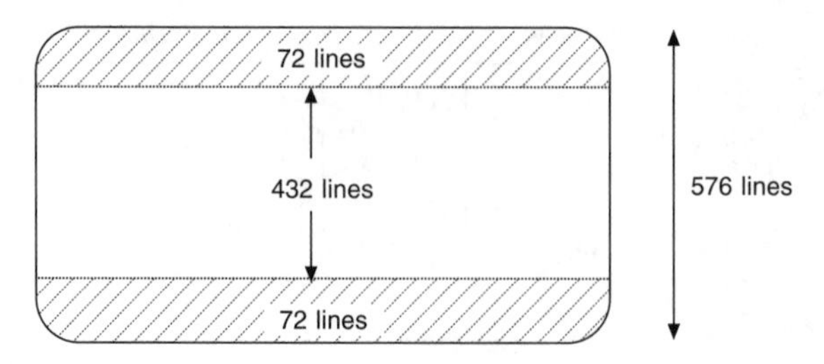

Figure 3.11
Letterbox display

bottom. In PAL video, the vertical resolution is reduced from 576 lines to 432

4. Widescreen, where the recording is in widescreen and the TV used for display is also widescreen.

Films with a wide format are converted to an anamorphic video source by squeezing the frame horizontally to fit a 4 : 3 format. At playback the shape of the displayed picture will depend on which mode is selected by the user, as illustrated in Figure 3.12. For a full-screen image the 16 : 9 mode must be selected, and the picture is then 'un-squeezed' back to its original proportions and, provided the widescreen television receiver is used, the full 16 : 9 image is displayed. With a standard 4 : 3 television receiver, the user may select a horizontally squeezed image with everything looking tall and thin, letterbox, or pan & scan. For the player to correctly pan & scan the anamorphic picture, vectors must be included in the MPEG-2 data stream recorded on the disc.

Telecine

There is another problem with encoding film images into DVD pictures, in that a cine film is shot at 24 frames per second while NTSC video uses 60 and PAL video 50 fields per second (i.e. 25 and 30 pictures per second respectively). A process of conversion has to take place, known as *telecine*. In the case of converting film to NTSC video a process known as *2–3 pull down* is used, in which one cine frame is converted into two fields and the following

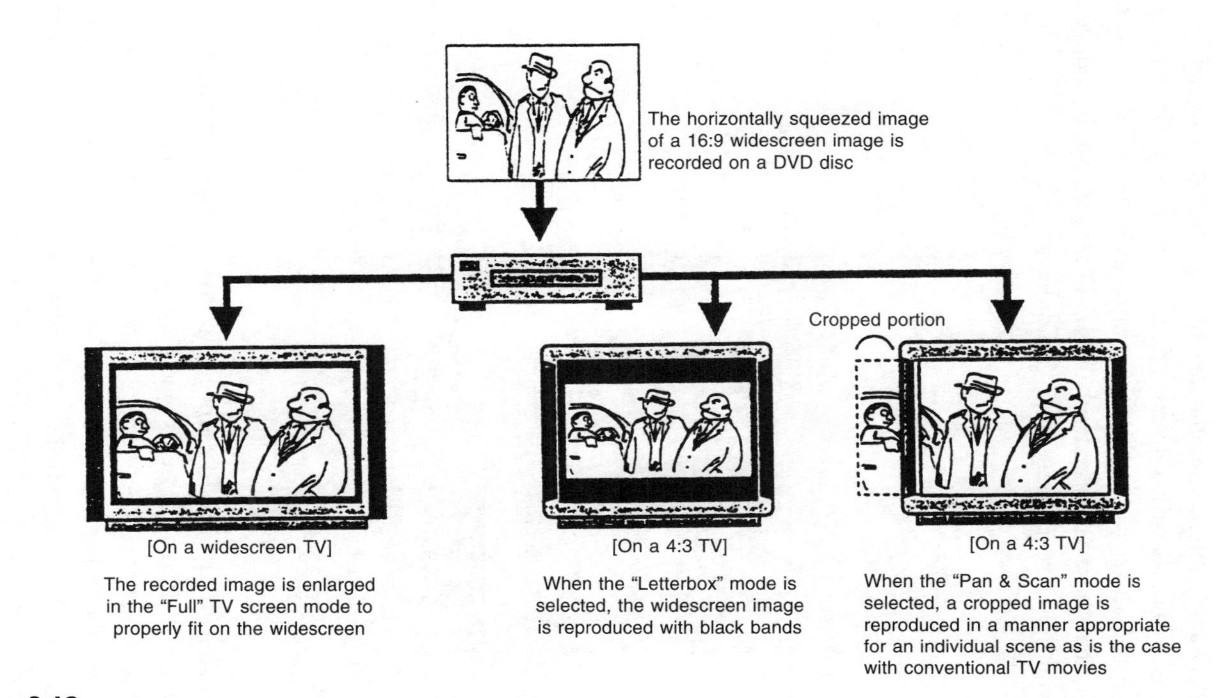

Figure 3.12
Multiple aspect ratio feature

is converted into three fields (see Figure 3.13). An extra field is thus created every two cine frames. This extra field is a repetition of the preceding field, and therefore does not have to be encoded for DVD applications; the encoder recognizes the extra field and removes it. This is called inverse telecine. The encoder then places appropriate flags in the video stream to indicate which fields to repeat, and these flags instruct the DVD player to reintroduce a fifth field, thus recreating the 2–3 pulldown.

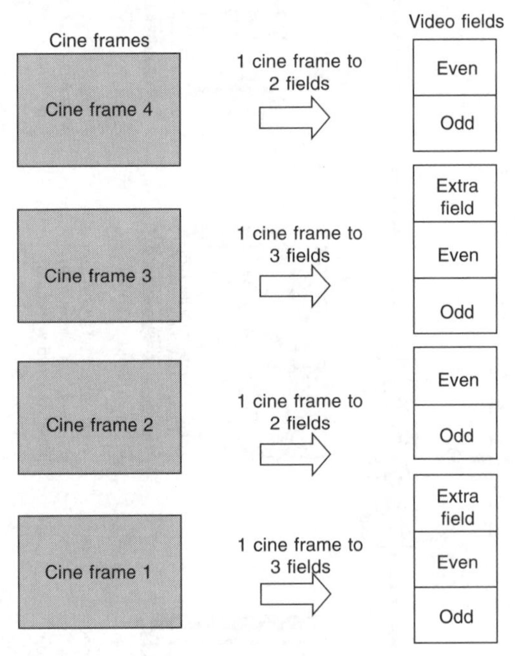

Figure 3.13
NTSC 2:3 conversion

When converting cine film images to PAL video each cine frame is converted into two fields, yielding 48 fields per second – a process sometimes known as *2–2 conversion* (Figure 3.14). At playback, the video is played at the slightly faster rate of 50 fields per second. A movie will last 2 minutes and 24 seconds less for every hour of film, and to compensate for this the film is run 4 per cent faster.

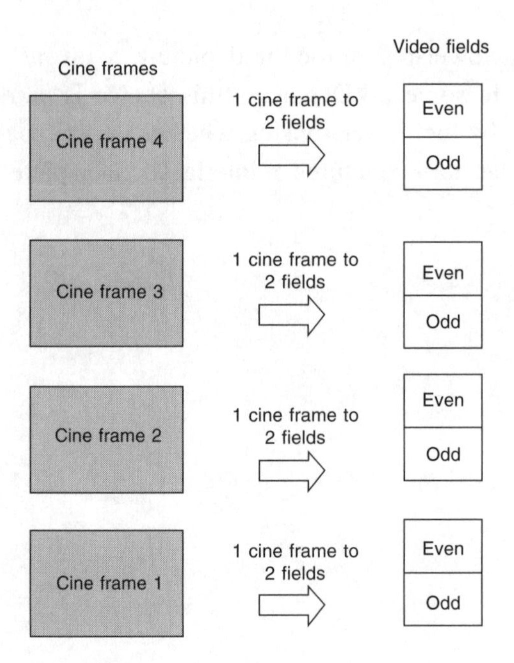

Figure 3.14
PAL 2-2 conversion

The pause function

One of the functions of a DVD player is the ability to pause the playback. When pause is activated, playback stops and a still picture of the last frame/field is displayed. Still pictures can be displayed by storing signals of the immediately preceding field or frame in digital memory, and playing them back repeatedly. Generally, a still image created from a frame memory has double the resolution of a field memory picture. However, in cases where an image containing a great deal of movement is taken by an interlaced video camera, frame memory may not be better than field memory. This is because the interlaced fields do not have much correlation between them. During frame pause, if two fields of this type are combined then blurring will be observed. For a clear pause picture images from movies and animation are best, where each frame is a self-contained 'pause' picture. If the original picture is interlaced and containing a great deal of movement, it is better to display a lower-resolution field picture to avoid the annoying blurring. In some DVD players this choice is carried out automatically by the player sensing whether the original pictures are non-interlaced film pictures or interlaced video pictures and acting accordingly.

CHAPTER 4

DVD ENCODING

Introduction

DVD information has three main components: video, audio, and sub-picture (which includes subtitles, captions, menus, karaoke lyrics, etc.). These three components are stored in what is known as streams. A *stream* literally means a data flow, something akin to a track on a tape recorder. DVD-video supports the following groupings:

Video	1 stream
Audio	Up to 8 streams
Sub-picture	Up to 32 streams

The streams are encoded separately to form video, audio and sub-picture (SP) *packets* before they are multiplexed to form the *program stream* (Figure 4.1). The video stream is encoded using MPEG-2 data compression techniques. The audio streams may be used to carry up to eight separate audio channels that can be used for multi-channel high quality sound as well as multilingual sound tracks. The DVD producer has a choice of three types of audio encoding; linear pulse code modulation (PCM), Dolby Digital (also known as AC-3), and MPEG-2 audio. Two other formats are also available; DTS (Digital Theatre System), and SDDS (Sony Digital

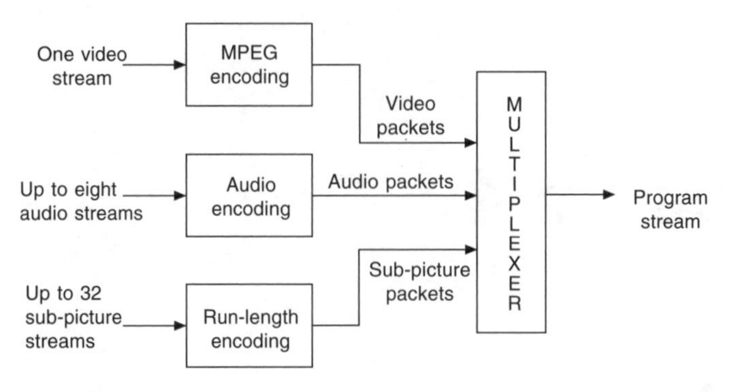

Figure 4.1
DVD streams

Distribution Sound). The sub-picture streams are used for displaying subtitles in different languages, menus, karaoke lyrics and information for interactive operations. Sub-pictures are encoded using bit mapping with run-length coding for data compression.

Each packet of data is fully identifiable in terms of data type, stream number and timing information for synchronization purposes. The packets are then multiplexed to form what is known as a program stream, which, following framing and the addition of data correction bits, it is stored on a disk. Unlike digital television broadcasting, in which the bit rate of the program stream is fixed and the data compression has to be adjusted to accommodate it, in DVD application a variable bit rate is used.

Video encoding

Video encoding consists of two major parts: video data preparation and video data compression (Figure 4.2). Video data preparation ensures that the raw coded samples of the picture frames are organized in a way that is suitable for data reduction. Video data compression is carried out in accordance with the internationally accepted standards established by the MPEG-2 system. Two major data reduction exercises are thus performed in the following order:

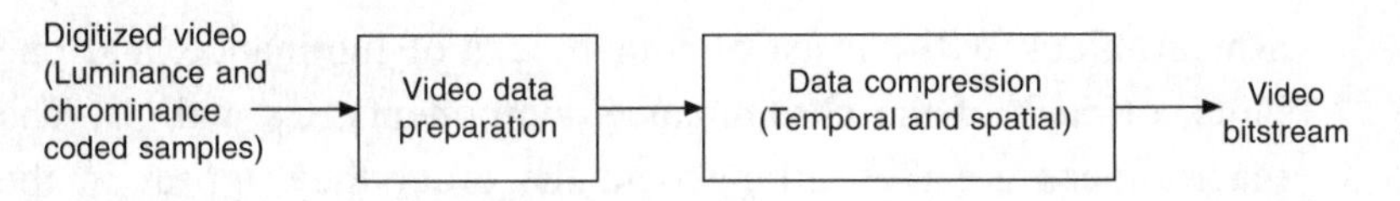

Figure 4.2

1. Temporal (time-related, i.e. frame-by-frame) redundancy removal
2. Spatial (space-related, i.e. within a frame) redundancy removal.

Temporal redundancy removal is an *inter-frame* data reduction technique that compares 16 × 16 pixel blocks of two successive picture frames, predicts the difference between them, and transmits a vector describing their movement, frame-by-frame. *Spatial* redundancy removal, known as *intra-frame* compression, removes unnecessary repetitions of the contents of an individual picture frame. It is carried out using a type of Fourier Transform known as *Discrete Cosine Transform* (*DCT*). The purpose of DCT is to transform sample values of an 8 × 8 block of pixels into coefficients. The number of coefficients is reduced by not transmitting the near-zero coefficients and by quantizing – i.e. rounding up or rounding down to a set of smaller number of integer values. Each coefficient is then translated into an 8-bit digital code, which forms the data bitstream.

Video data preparation

The video information enters the video encoder in the form of line-scanned coded samples of luminance Y and chrominance C_R and C_B. Video preparation involves regrouping these samples into 8 × 8 *blocks* to be used in spatial redundancy removal. Following that stage, coded samples are arranged into 16 × 16 *macroblocks* to be used in temporal redundancy removal. The macroblocks are then grouped into slices to be used for further data reduction techniques. The actual components making up a macroblock are determined by the MPEG-2 profile used. Using 4 : 2 : 0 sampling, a macroblock will consist of four blocks of luminance, and one block of each of the chrominance components, C_R and C_B. The macroblocks are then arranged in the order they appear in the picture to form a slice. The complete picture frame may then be reconstructed by a series of these slices.

Temporal data compression

Temporal compression exploits the fact that the difference between two successive picture frames is very slight, and thus it is not necessary to transmit the full contents of each picture frame because most of the contents are merely a repetition of the previous frame. Temporal compression is carried out on a *group of pictures* (*GOP*) composed of 12 non-interlaced frames. The first frame of the group is known as the *I-frame* (I for inter; Figure 4.3).

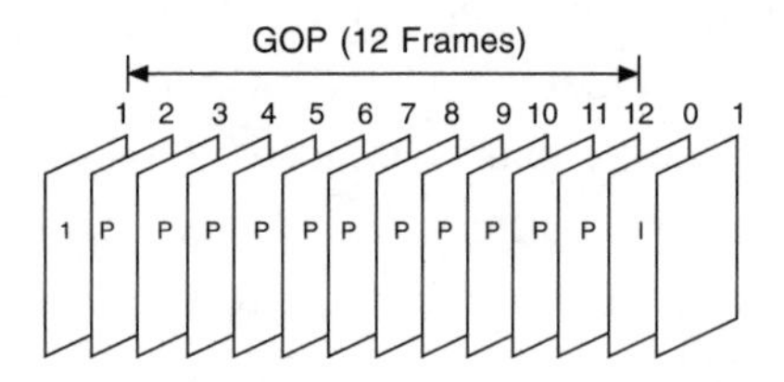

Figure 4.3

To illustrate the concept of temporal compression consider the sequence of two frames shown in Figure 4.4, each frame consisting of nine cells arranged in three lines. The I-frame, being the first frame, acts as a reference frame and must be sent without any data reduction. The contents of the cells are scanned, and are described as follows: *lion, horse, frog, globe, chair, bulb, leaves, tree, and traffic lights*. The second frame is slightly different from the first, and may be described in full by describing the contents of each cell in the same way as in the I-frame: *plane, horse, frog, globe, lion, bulb, leaves, tree, traffic lights.* However, such an exercise involves a repetition of most of the elements of the first frame, namely *horse, frog, globe, bulb, leaves, tree* and *traffic lights.* The repeated elements are said to be redundant, because they do not

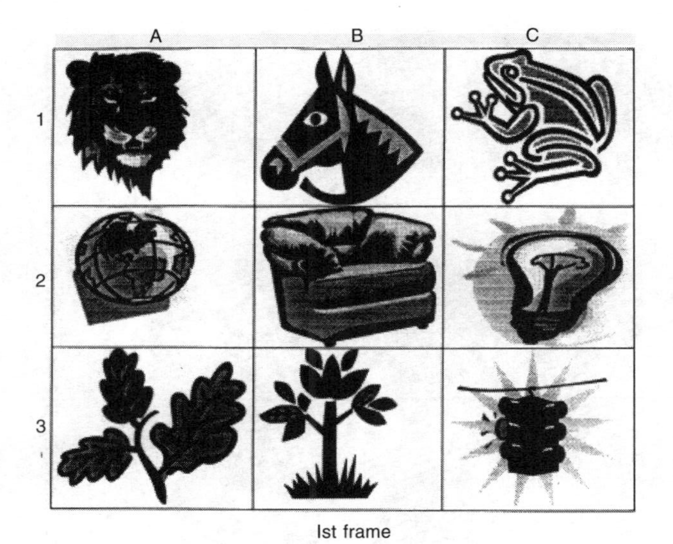

Ist frame

2nd frame

Figure 4.4

add anything new to the original composition of the frame. To avoid redundancy, only the changes in the contents of the picture are described. These changes may be defined by two aspects: the movement of the *lion* from cell A1 to cell B2, and the introduction of a *plane* in cell A1. The first of these is described by a *motion vector*. The newly introduced *plane* is derived by a slightly more complex method. First the motion vector is added to the first frame to produce a predicted frame (Figure 4.5), and the predicted picture is then subtracted from the second frame to produce a *picture difference* (Figure 4.6). Both components (motion vector and picture difference) are combined to form what is referred to as a *P-frame*. The process is then repeated, with the third frame compared with the previous P-frame to produce a second difference frame, and so on, until the end of the group of 12 picture frames. A new reference I-frame is then produced for the next group of 12 frames, etc. This type of prediction is known as *forward prediction*.

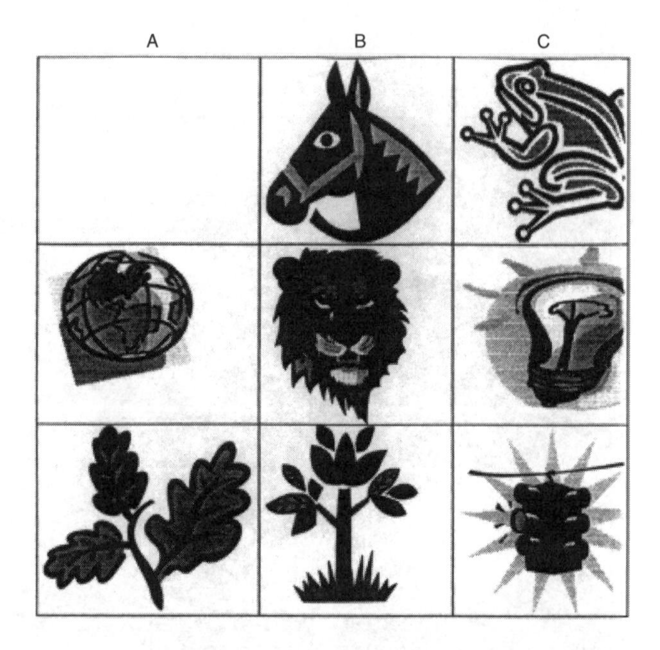

Figure 4.5
Predicted frame produced by the addition of the motion vector to the previous frame

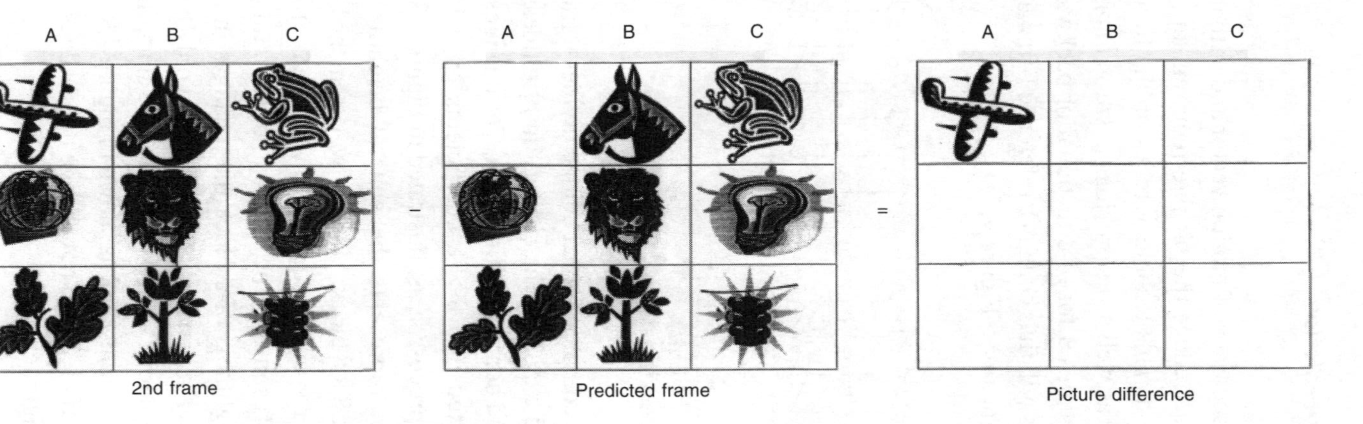

Figure 4.6

Bi-directional prediction

To improve the accuracy of motion vector and with it the predicted frame, and hence reduce the bit rate requirement, *bi-directional prediction* is used. This technique relies on the future position of a moving object as well as its previous position to produce what is known as a *B-frame* (B for bi-directional or backward). A typical group of pictures will thus contain a reference I-frame and a number of P- and B-frames (see Figure 4.7).

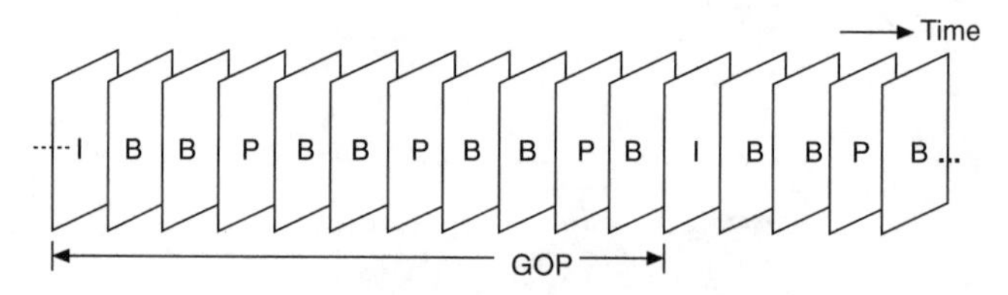

Figure 4.7

Spatial data compression

The spatial data compression technique divides a frame into 8 × 8 pixel blocks and examines the changes in brightness of the block in the vertical and horizontal directions. The changes in brightness are then converted into 8 × 8 coefficients by a DCT (discrete cosine transfer) processor, as illustrated in Figure 4.8. The top left-hand coefficient represents the general brightness or the DC component of the 8 × 8 block. Moving away from the DC component represents higher spatial frequencies, with the bottom right-hand coefficient representing the highest horizontal and vertical frequencies. Brightness of a block that changes in the horizontal direction only, e.g. a greyscale display would only have the values of the coefficients along the top horizontal line changing (decreasing as the brightness diminishes), as shown in Figure 4.9. In the vertical direction the coefficients do not change their values, since there is no change in brightness.

Discrete cosine transformation does not directly reduce the number

Original block

146.	144.	149.	153.	155.	155.	155.	155.
150.	151.	153.	156.	159.	156.	156.	156.
155.	155.	160.	163.	158.	156.	156.	156.
163.	161.	162.	160.	160.	159.	159.	159.
159.	160.	161.	162.	162.	155.	155.	155.
161.	161.	161.	161.	160.	157.	157.	157.
161.	162.	161.	163.	162.	157.	157.	157.
160.	162.	161.	161.	163.	158.	158.	158.

DCT processor

DCT block

314*	–0.26	–3	–1.30	0.53	–0.42	–0.68	0.33
–5.65	–4.37	–1.56	–0.79	–0.71	–0.02	0.11	–0.30
–2.74	–2.32	–0.39	0.38	0.05	–0.24	–0.14	–0.02
–1.77	–0.48	0.06	0.36	0.22	–0.02	–0.01	0.08
–0.16	0.21	0.37	0.39	0.03	–0.17	0.15	0.32
0.44	–0.05	0.41	–0.09	–0.19	0.37	0.26	–0.25
–0.32	–0.09	–0.08	–0.37	–0.12	0.43	0.27	–0.19
–0.46	0.39	–0.35	–0.46	0.47	0.30	–0.14	–0.11

Figure 4.8
DCT Processing
*DC component

of bits required to represent the 8 × 8 pixel block. Sixty-four pixel sample values are replaced by 64 DCT coefficient values. The reduction in the number of bits follows from the fact that for typical blocks of natural images, the distribution of the DCT coefficients is non-uniform. In an average DCT matrix most of the coefficients are concentrated at and around the top-left corner, with the bottom right quadrant having very few coefficients of any substantial value. Bit-rate reduction may thus be achieved by not transmitting the zero and the near-zero coefficients. Further bit reduction may be introduced by weighted quantizing, and by special coding techniques of the remaining coefficients. The greyscale display shown in Figure 4.9 has only four coefficient values and 60 zeros. Further simplification of the DCT coefficients is introduced by rounding the coefficient values up or down – a process known as quantization.

Zigzag scanning of the DCT matrix

Before coding the quantized coefficients the DCT matrix is reassembled into a serial stream by scanning each coefficient in the pattern shown in Figure 4.10, starting at the top-left cell (the average brightness component). The zigzag scan pattern makes it more likely that the significant-value coefficients will be scanned first. In this example, the scanned order is 315, 2, –4, –2, –3, –3, –2, –2, –3, 0, 2, 0, 0 and 1. No further transmissions are necessary since the remaining coefficients are zero and thus contain no information. This is indicated by a special *end of block* (EOB) code, which is appended to the end of the scan. In some cases, a significant coefficient may be trapped within a block of zeros. In such cases, other special codes are used to indicate a long string of zeros.

Coding of DCT coefficients

The coding of the quantized DCT coefficients employs two

Picture

223	191	159	128	98	72	39	16
223	191	159	128	98	72	39	16
223	191	159	128	98	72	39	16
223	191	159	128	98	72	39	16
223	191	159	128	98	72	39	16
223	191	159	128	98	72	39	16
223	191	159	128	98	72	39	16
223	191	159	128	98	72	39	16

Sample values

43.8	–40	0	–4.1	0	–1.1	0	0
0	0	0	0	0	0	0	0
0	0	0	0	0	0	0	0
0	0	0	0	0	0	0	0
0	0	0	0	0	0	0	0
0	0	0	0	0	0	0	0
0	0	0	0	0	0	0	0
0	0	0	0	0	0	0	0

DCT coefficients

Figure 4.9
DCT coefficients of a greyscale display

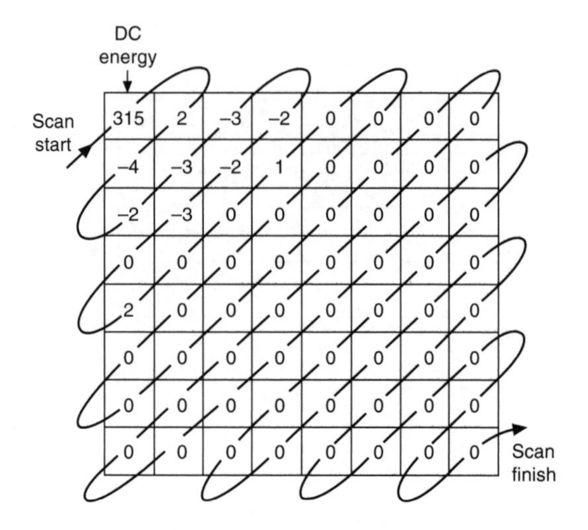

Figure 4.10

compression techniques; run-length coding (*RLC*) and variable length coding (*VLC*). Run-length coding exploits the fact that among the non-zero DCT coefficients there is likely to be a number of successive occurrences of zero coefficients. Instead of transmitting these coefficients as zeros, the number of zero coefficients is encoded as part of next non-zero coefficient. For instance, consider the following set of DCT values:

12, 6, 0, 4, 3, 0, 0, 5, 7, 0, 0, 0, 0, 0

RLC will form the series of DCT values into the following groups:

(12), (6), (0, 4), (3), (0, 0, 5), (7), (0, 0, 0, 0, 0)

The number of codes required to transmit these values has thus been reduced from 14 to 7 by grouping any zero coefficient or a run of zero coefficients together with the following non-zero coefficient – (0, 4) and (0, 0, 5). Each group is then given a unique code. The final run of zeros is grouped together and replaced by a single end of block (EOB) code.

The precise code used for each group is determined by the probability of its occurrence. Those occurring most frequently are given a shorter codeword than those that occur infrequently. This is the principle of *variable length coding*, also known as *entropy coding*. The most well known method for variable length coding is the Huffman code, which assumes previous knowledge of the probability of each DCT value. For instance, the DCT value 3, which occurs frequently, is allocated a 6-bit codeword – namely 0010 10 – while the infrequent DCT value 12 is allocated a 14-bit codeword – 0000 0000 1101 00 – and a 0 followed by a 4 is allocated the 9-bit codeword 0010 0001 0. End of block (EOB), which is the most frequently occurring string, is allocated a mere 2-bit codeword, namely 10.

The codewords are held in a look-up table in a read-only memory (ROM) chip. At the decoding stage of the DVD player, the bitstream has to be resolved into its original codewords. To make this possible, the Huffman algorithm ensures that short codewords never match the start of any other longer codeword – i.e. they must not be a prefix of longer codewords. For instance, since the codeword for EOB is 10, none of the other codewords can be prefixed with binary 10. The decoder can then resolve the codewords by examining increasing number of bits until a match with the look-up table stored in ROM is found. Having deduced the codeword, the actual DCT values may then be obtained by using another ROM look-up table.

Both RLC and VLC are known as *lossless* coding techniques. Lossless codes, as the name suggests, do not introduce any losses, and as such they are fully reversible at the decoding stage.

Variable bit rate

The bit rate is the volume of data that can be sent to be stored on the disk in a unit of time – i.e. bits per second (bps). The larger the rate, the larger the amount of data sent, which generally results in higher image quality. Following MPEG video compression, a bit rate of about 5 Mbits/s is required for PAL-standard video. If a *constant* bit rate (CBR) of 5 Mbits/s is used, fast-moving pictures that require a higher bit rate will suffer in quality. On the other hand, parts of the picture with virtually no motion require far fewer data, so high picture quality for these parts can be achieved with lower bit rate. If a constant bit rate is set, then fast-moving parts will be starved of data bits while slow-moving parts will have more bits than necessary to maintain quality. With a CBR, a lot of bits are wasted if the actual volume of data needed is small – as is the case in near-still parts of a picture. Furthermore, since the data capacity of the disc equals the bit rate multiplied by the recording time, for a given disc capacity a higher bit rate will result in a shorter recording time. For this reason, DVD-video invariably employs a variable bit-rate technology with a maximum speed of 9.8 Mbits/s. The bit rate is varied according to the condition of the picture; a high bit rate is used for the parts of the picture that require a large data volume (such as fast-moving cars) and a lower bit rate is used for parts of the picture that require a small data volume (such as a clear sky). At the encoding stage, a method known as *two-pass encoding* is employed to ensure the most efficient use of the bit rate available. The two-pass technique, as the name suggests, involves reading the original digital data twice. In the first pass, the pictures are analysed to determine what information each image requires and calculate the amount of data to be allocated to individual scenes without exceeding the maximum allowable bit rate. In the second pass, the information gathered in the first pass is implemented to produce data compression. In this way, an average of 3.5 Mbits/s is adequate for good quality PAL-standard pictures.

Audio encoding

DVD discs may incorporate up to eight audio streams, and each one can convey mono or multi-channel audio using one of three coding techniques:

1. MPEG-2 audio
2. Dolby Digital (also known as AC-3)
3. Linear pulse code modulation (PCM).

Both MPEG-2 and AC-3 coding involve audio compression to reduce the data bit rate. Linear PCM does not include any form of data compression, and thus it has the highest quality and the greatest bit-rate requirements.

Two other formats are also available as additional options: *DTS* (Digital Theatre System) and *SDDS* (Sony Digital Distribution Sound).

Multi-channel formats

The audio channels of a multi-channel system may be combined in several modes (Figure 4.11):

Mode	Channels
1/0	Mono
2/0	Stereo (Right, Left)
3/0	Right, Left, Centre
2/1	Right, Left, Surround
3/1	Right, Left, Centre, Surround
2/2	Right, Left, Right Surround, Left Surround
3/2, also known as 5	Right, Left, Centre, Right Surround, and Left Surround
5/2, also known as 7	(Supported by MPEG-2 only) Right, Left, Centre Right, Centre, Centre Left, Right Surround, Left Surround

In addition a low frequency effect (*LFE*), also known as a *subwoofer* channel, dedicated to frequencies of 120 Hz and less is available with all combinations – such as Dolby Digital 5.1, which adds a subwoofer to the five-channel 3/2 mode, and the 7.1, which adds a subwoofer to the seven-channel 5/2 mode. The LFE channel provides the theatre effect.

In order to convert a multi-channel recording to a simple stereo

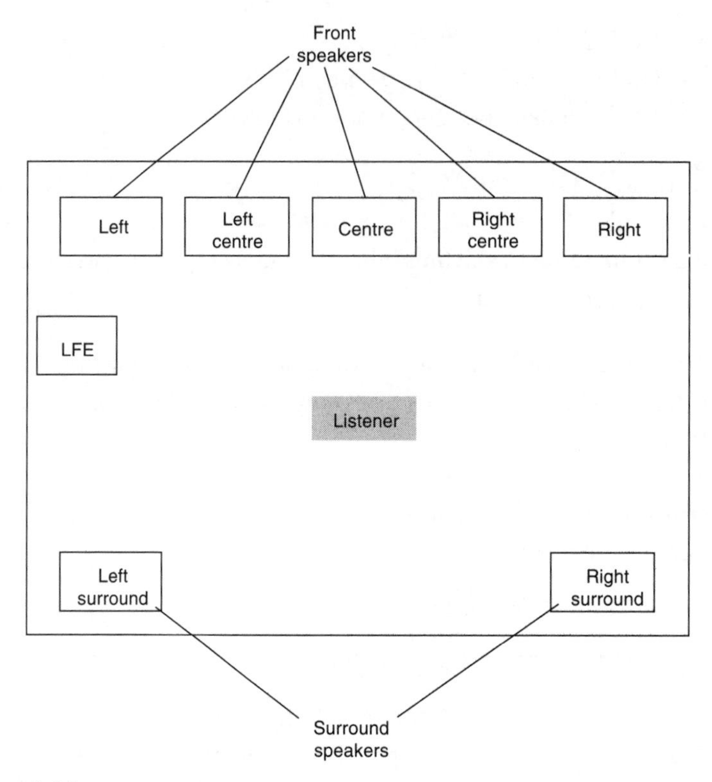

Figure 4.11

output, a process known as *downmixing* has to be performed. This involves matrixing the centre and surround channels onto the main stereo channels in accordance with a defined mathematical formula.

MPEG audio encoding

In DVD applications, audio signals are sampled at a frequency of 48 kHz at 16 bits. This is then followed by audio data compression. MPEG audio standards define three coding layers, which offer different compression rates for a given audio quality: layers I, II and III. Layer I has a smallest compression ratio, and therefore needs a comparatively high bit rate of 192 kbits/s per channel (or 384 kbits/s for two-channel stereophonic sound). Layer II has a higher level of data compression. For an equivalent audio quality, layer II requires a bit rate that is 30–50 per cent smaller than that required by layer I, at the expense of a moderately more complex encoder and decoder. The bit-rate requirement for layer II is 128 kbits/s per channel (256 kbits/s for stereo). MPEG layer III, more commonly known nowadays as *MP3*, supports a compression rate twice that of layer II, but the encoder and decoder are substantially more complex. Using layer III encoding, hi-fi quality sound requires a mere 64 kbits/s per channel (128 kbits/s for stereo). The combination of CD quality sound and a very low bit rate makes MP3 so useful in downloading audio pieces across the Internet. DVD specification supports layers I and II only. Nonetheless, because of the increasing popularity of MP3, most modern DVD players provide MP3 audio-decoding capability.

Before data compression takes place, the audio information is divided into fixed-size blocks or sub-bands of audio samples: 384 samples at layer I audio encoding, 1152 samples divided into three 384-sample blocks at layer II encoding. Layer III encoding employs a more complex system of sample block processing. MPEG audio compression is carried out on each sub-band separately, and is achieved by the use of special algorithms that remove parts of the audio signals without affecting the sound quality – a process known as masking. *Masking* makes use of a psycho-acoustical model that has been developed following research including a large number of people.

Masking

It is well known that the human ear can perceive sound frequencies in the range of 15 Hz to 20 kHz. Furthermore, the sensitivity of the human ear is not linear over the audio frequency range. Experiments show that the human ear has a maximum sensitivity over the range of 1–5 kHz, and that outside this range its sensitivity decreases as shown in Figure 4.12. The curve represents the hearing threshold, with the sounds below the curve not being perceived by the human ear and thus not needing to be transmitted. The hearing threshold will, however, change in the presence of multiple audio signals,

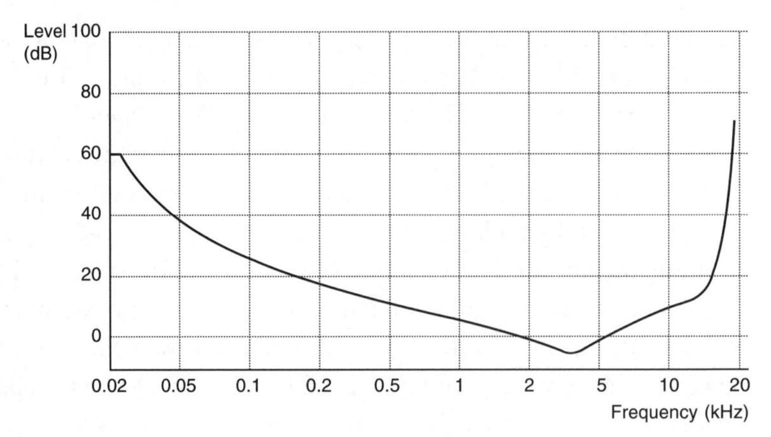

Figure 4.12
Graph showing the sensitivity of the human ear to audio frequencies

which could '*mask*' the presence of one or more sound signals. There are two types of audio masking; *spectral* (or frequency) masking, where the signals occur simultaneously, and *temporal* (i.e. time related) masking, where the signals occur in near time proximity to each other. Spectral masking exploits the particular characteristics of the human hearing, where a quiet sound is made inaudible by a loud sound of near frequency, and quieter higher frequencies are masked by louder lower ones. Temporal masking exploits the fact that a sound of high amplitude will tend to mask sounds immediately preceding it (pre-masking) and immediately following it (post-masking). Temporal masking represents the fact that the ear has a finite time resolution, whereby sounds arriving over a period of about 30 ms are averaged to produce a louder sensation whereas sounds arriving outside that time period are temporally discriminated and are perceived separately. It follows that what the human ear hears at any one time depends on a number of variables, including the range and loudness of the sound as well as the range and loudness of any sounds that have preceded or succeeded it. Research has defined a masking curve as a basis for what is known as a *psycho-acoustical model* of the human ear. This model is used as the basis of the MPEG audio encoder to determine which audio frequencies may be discarded as redundant for any given combination of sound frequencies.

Additional audio channels, such as the LFE subwoofer, are placed on separate extension streams.

MPEG-2 audio encoding is allocated a variable bit rate of 64 to 912, with 384 kbits/s being the normal average rate. In most application, the audio bit rate is fixed at the average value.

Dolby Digital (AC-3) encoding

Dolby Digital, also known as AC-3 (AC for audio compression), supports all channel formats up to 5.1. It uses a sampling rate of 48 kHz, with an average of 16 bits allocated to each sample. Its compression technique differentiates between short transient signals and long continuous sounds, giving prominence to the latter in the form of long sample blocks compared to those of the transient sounds. AC-3 uses a frequency transform technique similar to the DCT employed in MPEG video encoding. It provides smoother encoding compared with MPEG, which creates arbitrary boundaries by its sub-band technique. Up to 24 bits per sample are allocated dynamically to compensate for different listening environments – e.g. theatre, home or auditorium. The bit rate may be variable although, as with MPEG audio encoding, a fixed rate is normally used. Bit rates of 384 or 448 kbits/s are typical.

The Dolby Digital system is a high quality digital sound system equipped with a channel dedicated to subwoofer output (LFE) in order to reproduce low frequencies below 120 Hz in addition to the other five channels. Unlike the conventional two-track Dolby Pro Logic system, Dolby Digital is a 5.1 track system with each channel discretely and digitally processed from the beginning, so they are independent, and this contributes to excellent channel separation. All five channels are reproduced at full bandwidths of 3 Hz to 20 kHz. With these features sound may be accurately reproduced in a home theatre context, with the intended positioning of sound image and feeling along with surround sound of the presence and power comparable to that of a movie theatre experience.

Linear PCM

Linear PCM (pulse code modulation) is uncompressed and is thus a lossless digital audio, which has been used in CD and most studio masters. Analogue audio is sampled at 48 kHz or 96 kHz, with 16, 20 or 24 bits per sample. (Audio CD is limited to a sampling rate of 44.1 KHz.) As a result of the absence of compression and the high rate of sampling and quantization, the bit rate could be excessively high. For this reason it is limited to 6.144 Mbits/s. The equivalent bit rates for MPEG and Dolby Digital are 448 and 384 kbits/s. Linear PCM supports up to eight channels; however, owing to the limit of 6.144 Mbits/s, for five or more channels the lower sampling rate of 48 kHz must be used (see Table 4.1).

Table 4.1 Linear PCM – sampling rates and maximum number of channels supported

Sampling rate (kHz)	*Quantization (bits/sample)*	*Maximum number of channels*
48	16	8
48	20	6
48	24	5
96	16	4
96	20	3
96	24	2

The bit rate required by the non-compressed linear PCM is normally too high to be accommodated by a DVD program, which may include other elements such as video and other audio configurations, within the limits imposed by DVD specification on the maximum bit rate available for the various elements (namely 9.8 Mbits/s). For this reason, *meridian lossless packing* (*MLP*) is used. MLP compresses data bit by bit, removing redundant data without any loss to quality. It achieves a compression ratio of 2 : 1 by using a combination of techniques – lossless matrixing, lossless waveform prediction and entropy coding.

Table 4.2 compares the audio bit rates for three audio configurations.

Table 4.2 Comparison between the audio bit rates for three audio configurations

AC03 (+0.1 channel)		Linear PCM							MPEG (+0.1 channel)	
Sampling frequency – 48 kHz			Sampling frequency – 48 kHz			Sampling frequency – 96 kHz			Sampling frequency – 48 kHz	
Quantization channels	Bit rate(kbits/s) (compressed)	Quantization	16 bit	20 bit	24 bit	16 bit	20 bit	24 bit	Quantization channels	Bit rate (kbits/s) (compressed)
1/0	64–448	Channel 1	768	960	1152	1536	1920	2304	1/0	64–192
2/0	128–384	Channel 2	1536	1920	2304	3072	3840	4608	2/0	128–384
3/0 or 2/2	192–448	Channel 3	2304	2880	3456	4608	5760		3/0 or 2/1	256–912
3/1 or 2/2	256–448	Channel 4	3072	3480	4608	6144			3/1 or 2/2	256–912
3.2	320–448	Channel 5	3840	4800	5760				3/2	320–912
		Channel 6	4608	5700					5/2	448–912
		Channel 7	5376							
		Channel 8	6144							

Multiple-language dubbing

DVD-video is capable of recording up to eight streams of sound with Dolby Digital, MPEG-2 or PCM. Multiple-language sound track is produced by dubbing each stream with a different language. At playback, a particular stream may thus be selected.

The sub-picture stream

While moving pictures and audio is the main business of a DVD-video, sub-pictures form an integral part of DVD production. Sub-pictures are bitmapped and compressed using run-length coding overlayed onto still or moving video. They may be simple menus providing choices for the user, or synchronized text such as lyrics in karaoke applications.

Bitmapping is a process of encoding where the intensity and colour of each pixel in the sub-picture is coded individually in the same way as with computer graphics. In the case of DVD menus, they are often handled by software designed for computer-based multimedia applications. In incorporating computer-based menu designs, it is essential to take account of the difference between the computer platform and the DVD platform. For computer displays pixels are '*square*' – hence a resolution of 640 × 480 or 720 × 540 would yield a 4 : 3 image on a computer screen. For TV applications pixels are '*rectangular*' – hence a 720 × 576 (PAL) and a 720 × 480 (NTSC) image would both produce a 4 : 3 aspect ratio on a television. Menus designed on a computer with a resolution of 720 × 540 (aspect ratio 4 : 3) must therefore be scaled up to 720 × 576 (PAL) or down to 720 × 480 (NTSC) to produce an undistorted display on a television screen. Furthermore, over-scanning in the television application renders 5–10 per cent of the image invisible – something that computer images do not suffer from – and this must be taken into account.

CHAPTER 5

FRAMING AND FORWARD ERROR CORRECTION

Introduction

DVD-video supports three main streams: video, audio and sub-picture. To these are added two other streams, one for presentation control information (*PCI*) and a second for data search information (*DSI*), as illustrated in Figure 5.1. Presentation control and data search provide the necessary information for navigation and search facilities. Before these streams are multiplexed, they are broken into 2-KB (2048 bytes) chunks of data. Header and error correction bits are then added to each chunk, resulting in 2046-byte video, audio, etc. PES packets. The header includes ID bits to distinguish between the various types of packets. There are thus five different types of packets: video, audio, sub-picture, PCI and DSI. Furthermore, DVD-video supports eight different types of audio packets for multi-channel and multilingual applications, and 32 types of sub-picture packets for multilingual subtitles, menus, karaoke, etc. The packets are then fed into the multiplexer to form the program stream, and this is then followed by framing and forward error correction to produce 2418-byte recording sectors. Before writing the data onto the disc, it is fed into an 8-to-16 modulator (EFM+), which doubles the recording sector size to 4836 bytes.

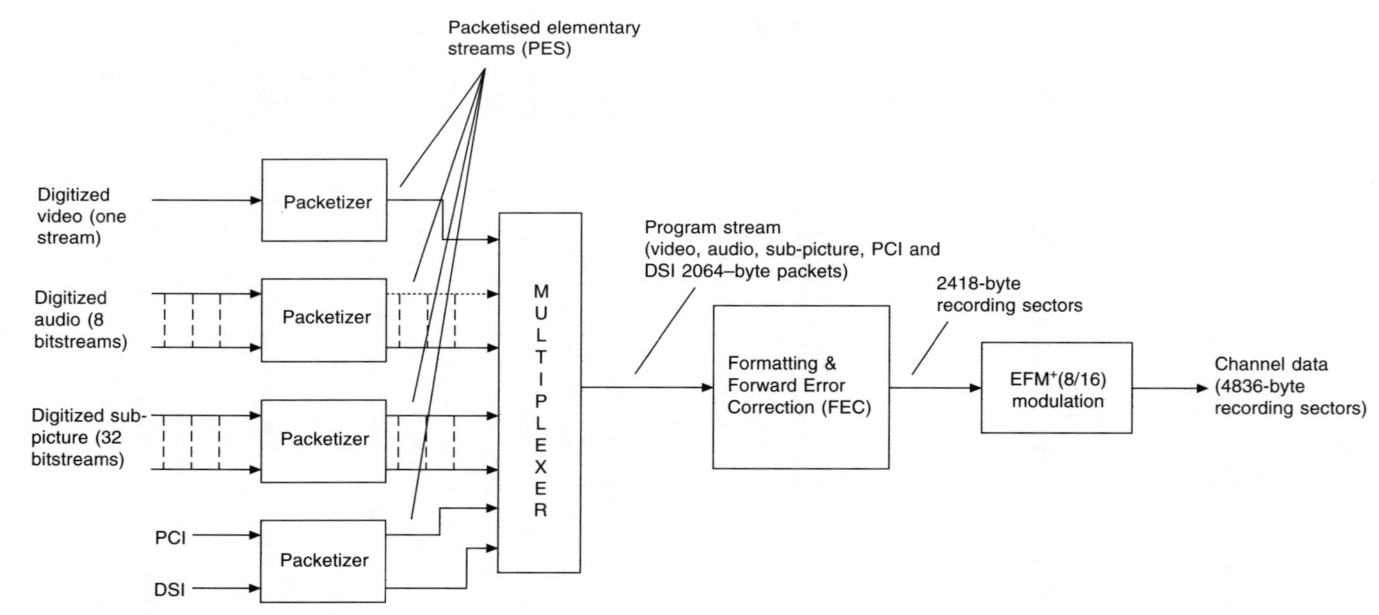
Packetised elementary streams (PES)
Digitized video (one stream)
Digitized audio (8 bitstreams)
Digitized sub-picture (32 bitstreams)
PCI
DSI
Packetizer
Packetizer
Packetizer
Packetizer
MULTIPLEXER
Program stream (video, audio, sub-picture, PCI and DSI 2064–byte packets)
Formatting & Forward Error Correction (FEC)
2418-byte recording sectors
EFM+(8/16) modulation
Channel data (4836-byte recording sectors)

Figure 5.1

PES packet construction

A PES packet (Figure 5.2) contains 2 KB of user data such as video, audio, and sub-picture information. The user data are preceded by a header and followed by a 4-byte error detection code (*EDC*) to form a packet, also known as a sector. However, before the header and the EDC bytes are added, the 2-KB user data bits are scrambled. Scrambling avoids long strings of zeros, to ensure that energy is evenly spread.

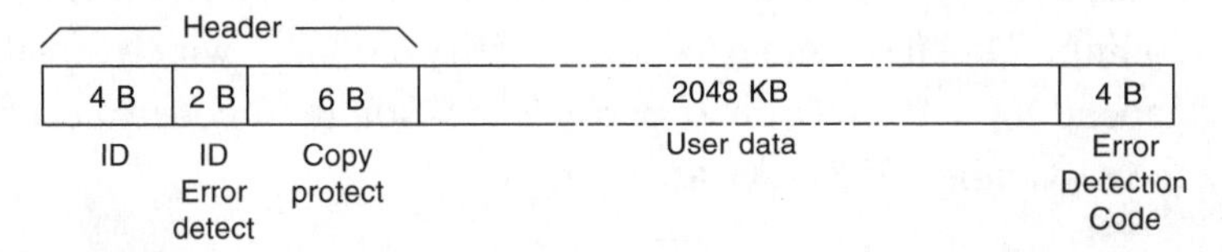

Figure 5.2

The header consists of three sections: a 4-byte ID section (consisting of a 3-byte sector number and a 2 bytes of error detection for the sector number) and 6-byte copy protection information.

Forward error correction

Digital signals, especially those with a high level of compressions, require an efficient error correction capability. Apart from normal errors of propagation and processing, errors may occur as a result of dust or of physical damage to the surface of the disc. To avoid such errors causing video or audio break-up or disruption, powerful forward error correction (*FEC*) techniques are employed that can correct a burst error length of approximately 2800 bytes, corresponding to physical damage of up to 6.0 mm. In comparison, the error correction technique used in audio CDs can correct a burst of errors of approximately 500 bytes, corresponding to 2.4 mm of physical damage.

Before error correction bits are added, the 2064 bytes of each data sector are arranged into 12 rows of 172 bytes (Figure 5.3). Sixteen data sectors (a total of 16 × 2064 = 33 024 bytes) are then grouped together to form an error code block (ECB), with each ECB comprising 12 × 16 = 192 rows of 172 bytes, an array of 192 × 172 = 30 024 bytes. At this stage, the 192 rows are interleaved in order to spread the rows and help in subsequent correction of error bursts. Error correction code (ECC) bits are applied to the block using Reed-Solomon block code (208, 192, 17) × (182, 172, 11). This involves adding a 16-byte outer parity code to each of the 172 columns, forming 16 new rows at the bottom and increasing the number of rows to 192 + 16 = 208. This is known as outer parity coding. The next stage is inner parity coding, which involves appending a 10-byte outer parity Reed-Solomon code to each of the re-vamped 208 rows as illustrated.

The final restructuring involves selecting 12 rows from the 182 byte × 192 block and adding one from the 16-row outer parity block to construct a 182 byte × 13 rows block (see Figure 5.4) to form one recording sector. Each recording sector thus consists of 12 + 1 = 13 rows of 182 bytes. The recording sector is then split up in the middle and a 2-byte sync inserted at the start of each half row, as in Figure 5.4. This creates a sector size of 93 × 2 × 13 = 2418 bytes, which is doubled to 4836 bytes following 8/16 modulation before it is written onto the disc row by row. This resulting bitstream is known as the *channel data.*

Non-return-to-zero encoding

Data are written on the disk using a non-return-to-zero inverted (NRZI) format, in which a transition from pit to land represents a one and a non-transition represents a zero. The effect of this is to halve the number of transitions required to represent the bitstream. Consider the stream of 10 pulses in Figure 5.5. With non-return-to-zero inverted, a change from land to pit or *vice versa* occurs only when a logic state 1 is followed by a 1, or a logic 0 is followed by a logic 1 (see Figure 5.5b). Otherwise, no change

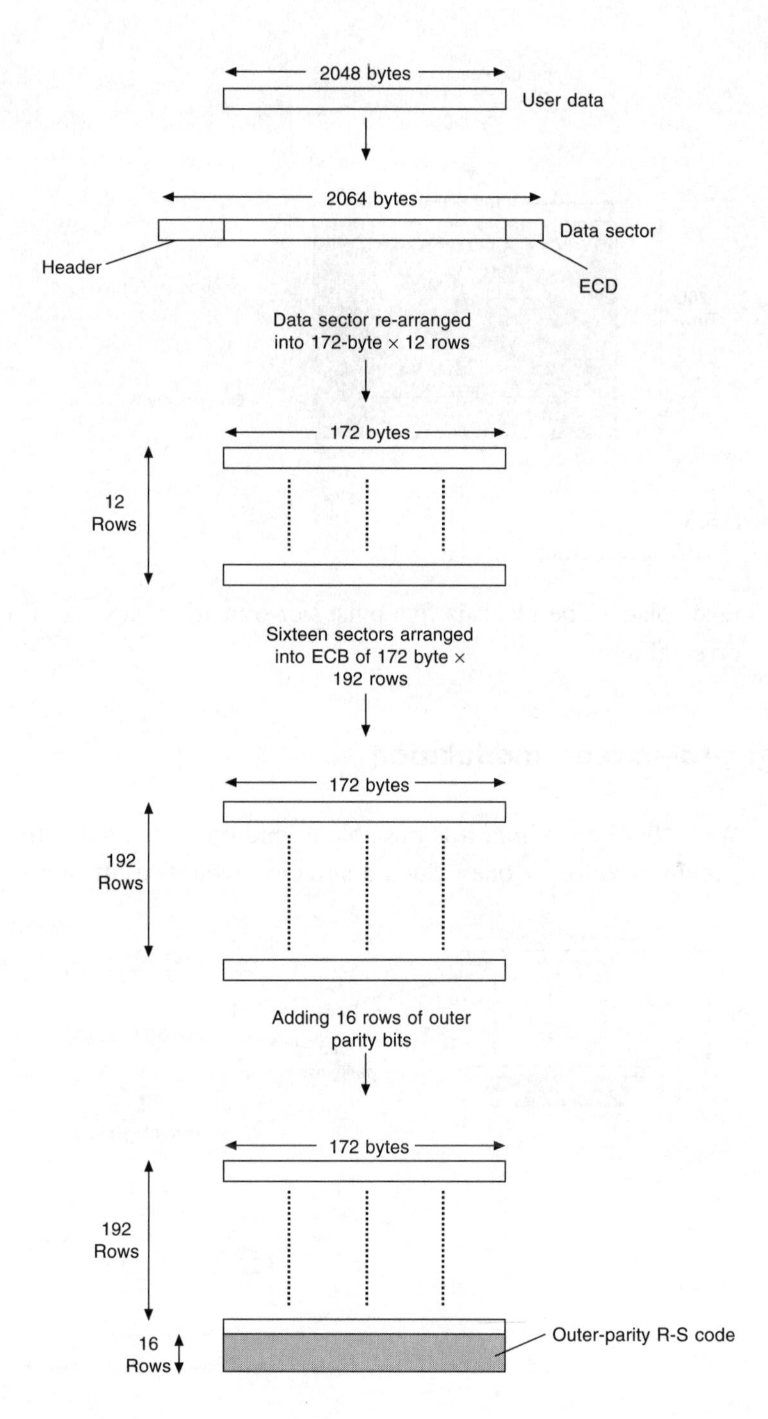

Figure 5.3 ***(Contd)***

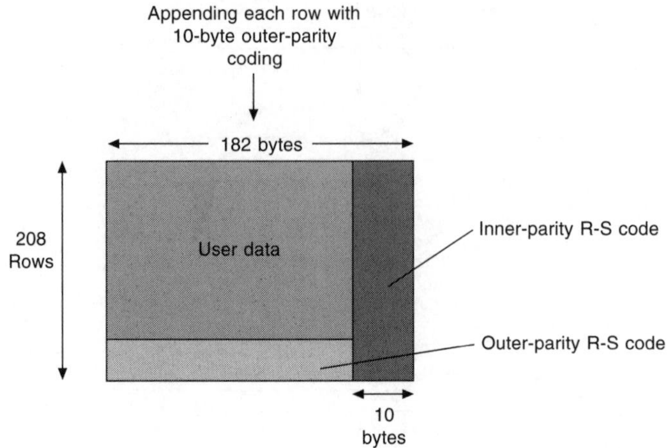

Figure 5.3
Data recording structure

takes place. The result is five pulses or transitions instead of the original ten.

Eight-to-sixteen modulation

With NRZI encoding, it is possible to end up with a long-stretch stream of zeros or ones. Such a situation would result in a long

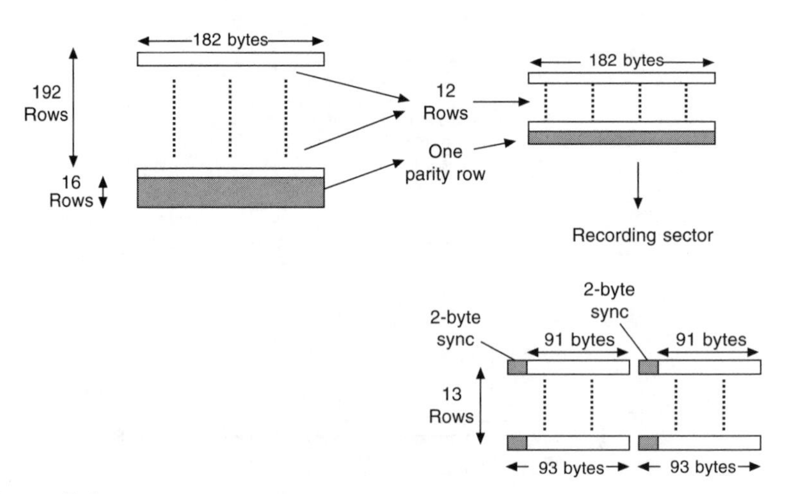

Figure 5.4
The recording sector

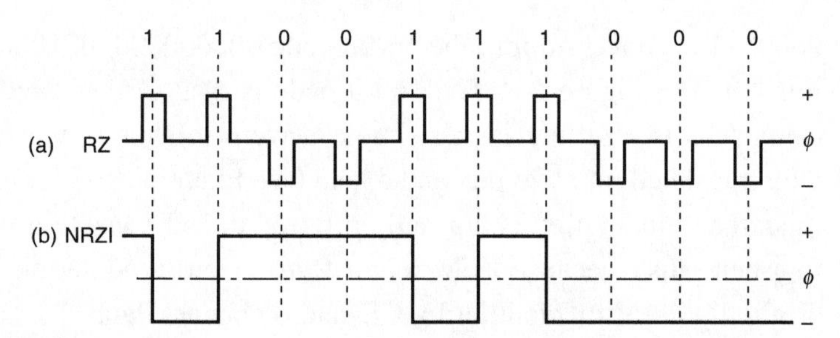

Figure 5.5
Non-return-to-zero halves the number of transitions

stretch of a constant high or low voltage with no transitional edge. Since clock synchronization depends on the regular occurrence of such a transitional edge, a long absence of such an edge would cause a breakdown of time synchronization at the player. To avoid this, 8/16 modulation is used.

Eight-to-sixteen modulation converts each word of 8 bits (1 byte) of data into a 16-bit code selected from a conversion table. An 8-bit word has $2^8 = 256$ different combinations or words. A 16-bit code on the other hand, has $2^{16} = 65\ 536$ different codes. Out of these 65 536 codes, 256 are carefully selected to minimize DC energy and reduce frequency. The selected codes are chosen in such a way that in each code there are at least two and at most ten zeros between any groups of ones. For example, the 8-bit word

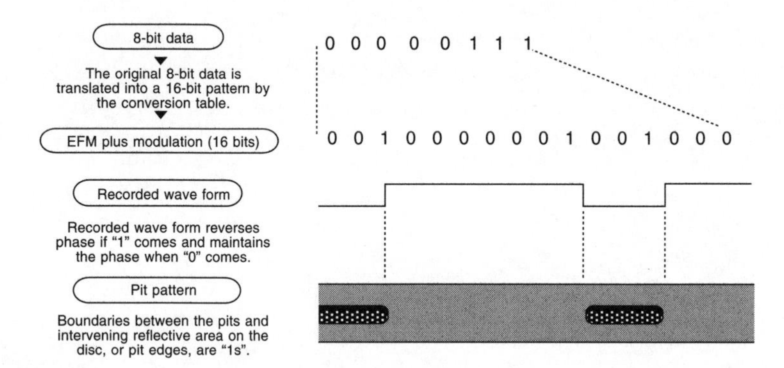

Figure 5.6
8–16 Modulation and pit pattern

0000011 is converted into the 16-bit code 0010000001001000, as illustrated in Figure 5.6. The 16-bit code is then further encoded using NRZI, resulting in the simple waveform shown, which is then translated into two pits and a land (see Figure 5.6). Eighteen-to-sixteen modulation is normally referred to as EFM+ (eight-to-fourteen plus), because EFM (eight-to-fourteen) modulation was originally used for modulating CD audio channel data.

CHAPTER 6

THE OPTICAL PICKUP UNIT

Introduction

The optical pickup unit (*OPU*) is mounted on an arm placed under the disc, along which the pickup head can move to follow a track or read different tracks. A DVD pick-up head assembly together with driving mechanism is illustrated on page vii.

Figure 6.1 shows a simplified diagram of a DVD player. The optical pickup unit, also known as the *optical head*, extracts data from the DVD disc using a *laser* beam, and then converts the reflected laser beam into an electric waveform known as the RF (radio frequency)

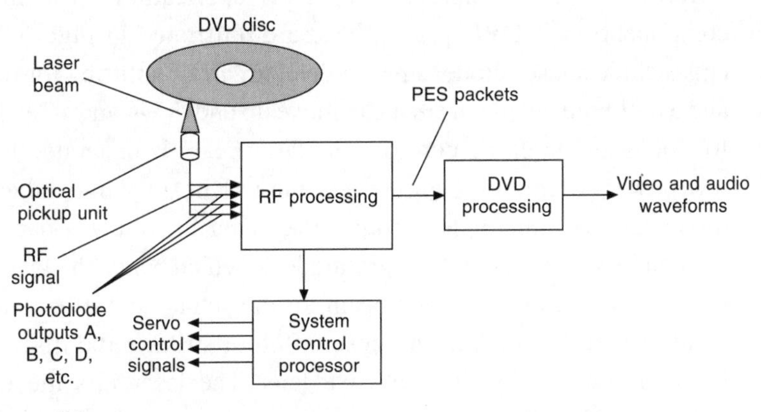

Figure 6.1

or HF (high frequency) signal. The RF signal is an analogue signal which, following conversion to a digital format and processing, is fed into the DVD processing unit to reproduce the original video, audio and other control and search information. The optical pickup unit also provides a number of other outputs – A, B, C, D, etc. – to be used to ensure accurate focusing, tracking and speed control of the DVD disc. The outputs from the pickup head are then fed into the RF amplifier and processor, which carry out the necessary processing of the RF signal to reproduce the original video, audio and other PES packets, as well as generating the necessary focus, tracking and other control signals.

The optical unit is one of the most critical parts of a DVD player. It combines the laser diode (which generates the laser beam), the photodiode array to detect the reflected laser beam, and all the necessary lenses and mirrors, in a single integrated device. It generates the laser beam, ensures that the beam focuses on and follows the recording tracks, detects the reflected waveform, and produces streams of data for processing and control by the DVD player.

The optical pickup unit is therefore a complex self-contained unit which carries out the function of reading or extracting the data from the surface of the disc. The construction of the OPU may be different for CD compared with DVD applications. The main components of a DVD pickup head are illustrated in Figure 6.2, and include a laser diode, a photo-detector array, splitting mirrors, and a coil-controlled lens that can move up and down and sideways for focus and tracking control. The laser beam is generated by a low power Aluminum Gallium Arsenic (AlGaAs) semiconductor diode. The collimator lens forces the beam to follow a parallel path on its way to the optical grating lens, which bends the laser to produce two beams – a main beam for the actual data stream, and a side beam for tracking purposes. Before striking the disc, the beam is focused by the objective lens. The laser hits the disc surface and is reflected back towards the objective lens. Provided

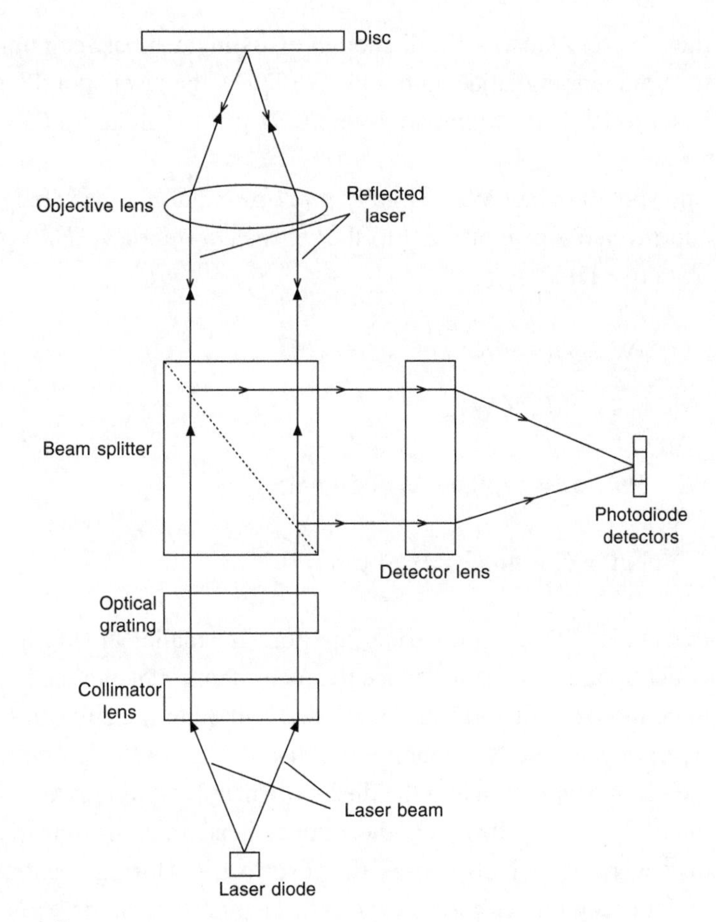

Figure 6.2
Optical pickup unit assembly (basic components)

the beam is accurately focused, the reflected beam returns along the same path as the incident beam towards the beam splitter. The beam splitter turns the reflected beam by 90° to direct it towards the photodiode detector assembly via the detector lens.

Focus depth and numerical aperture

With a track pitch as small as 0.74 μm and a minimum pit length of 0.40 μm (compared with 1.6 μm and 0.84 μm respectively for audio CDs), the *spot* size hitting the information surface of the disc must be very small – small enough to distinguish between pits and to avoid reading adjacent tracks. For DVD, the laser spot diameter is set to 0.54 μm (compared with 0.87 μm for an audio CD).

The spot diameter W is directly proportional to the wavelength λ, and inversely proportional to the *numerical aperture* (NA) of the objective lens:

$$W = \lambda/(2 \times NA)$$

$$NA = \lambda/(2 \times W)$$

With W = 0.54 μm and λ = 650 nm,

$$NA = 650\ nm/2 \times 0.54\ \mu m = 0.6$$

For audio CD, the numerical aperture is smaller at 0.45 μ. The effect of larger NA is to reduce the focus depth, causing the readout to be more sensitive to disc thickness and other irregularities. The first is overcome by reducing the disc thickness (to 0.6 mm) and the second by increasing the disc's stiffness by gluing two 0.6 mm plates together. However, the optical head remains sensitive to disc warping, which causes the disc to tilt. During readout the optical head moves across the disc; a changing tilt angle will cause what is known as a *skew error*, which will disturb the reading of the data. For this reason a tilt or skew sensor may be mounted on the surface of the optical unit to produce a skew error signal, which is used to ensure that the optical head moves in parallel to the disc surface.

Table 6.1 provides a comparison between DVD and CD parameters.

The photodiode detector assembly

The photodiode assembly consists of a number of photodiodes that detect the strength of the reflected laser beam and produce a

Table 6.1 Comparison between DVD and CD parameters

Parameter	*CD*	*DVD*
Track pitch	1.6 µm	0.74 µm (740 nm)
Minimum pit length	0.84 µm	SL: 0.40 µm DL: 0.44 µm
Linear velocity	1.21 m/s	3.49 m/s
Channel bit rate	4.3218 Mbits/s	26.16 Mbits/s
Objective lens	Plastic	Glass
Numeric aperture (NA)	0.45	0.6
Wavelength (λ)	780 nm	650 nm
Spot diameter	0.87 mm	0.54 mm

number of signals, including the RF, focus and tracking error signals. The *RF signal* contains the actual video, audio and other information. Since the pit and lands reflect the beam with different strengths, the output level from the photodiodes will thus represent these as a bitstream of zeros and ones.

The detector that monitors the main beam is divided into four quadrants or segments – A, B, C and D (see Figure 6.3). The sum of the outputs of the four segments (A + B + C + D) represents the RF signal strength, which is used for data processing. The same four-quadrant photodiode is used to produce the focus error signal.

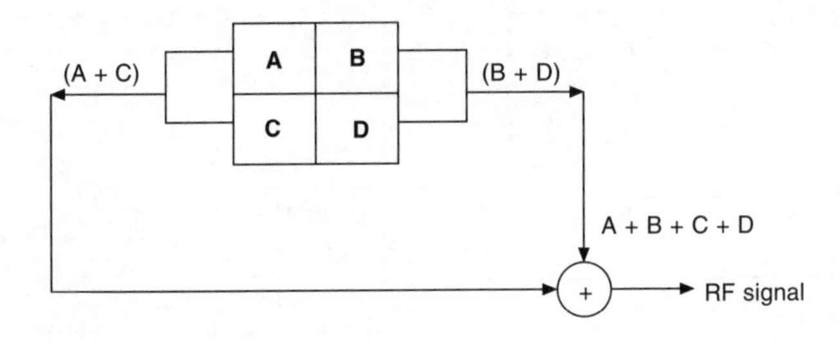

Figure 6.3
The photodiode

Focus error

When the beam is correctly focused, the reflected beam forms a circular pattern on the photodiode (see Figure 6.4a). However, if the beam is off focus, it forms an elliptical shape with a different aspect ratio. By comparing (A + C) with (B + D), a *focus error* (FE) signal is produced as shown in Figure 6.4 (b and c). When the beam is in focus (A + C) – (B + D) = 0. Otherwise, FE will be positive when the focus is too short and negative when the focus is too long.

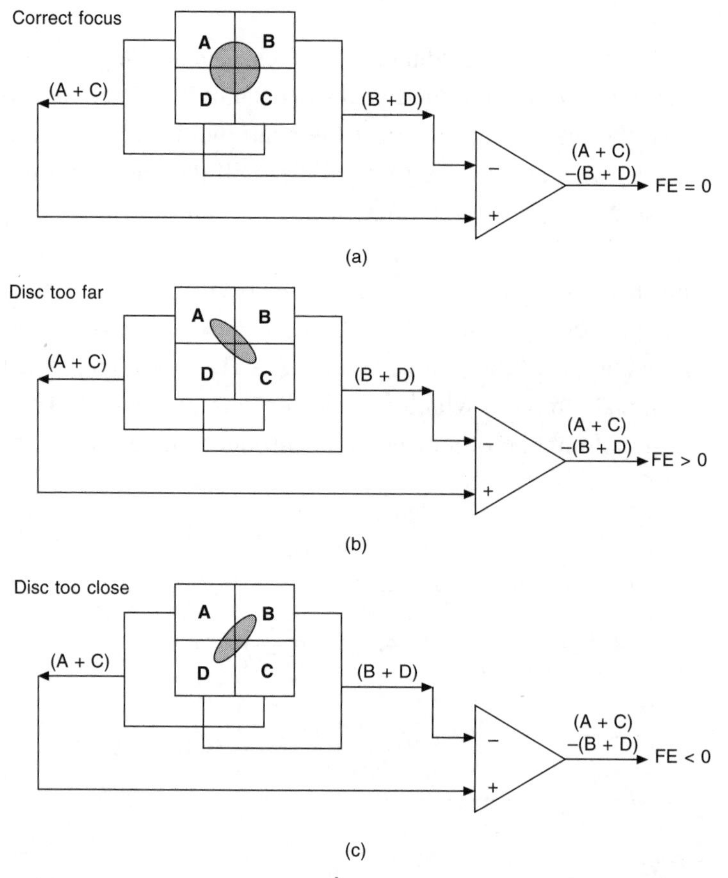

Figure 6.4

Tracking error

Apart from the main beam, which is used for the RF and FE signals, there are two side beams produced by the grating lens. These two side beams are directed to two separate photodetectors (E and F) placed on each side of the main beam photodiode assembly, as shown in Figure 6.5.

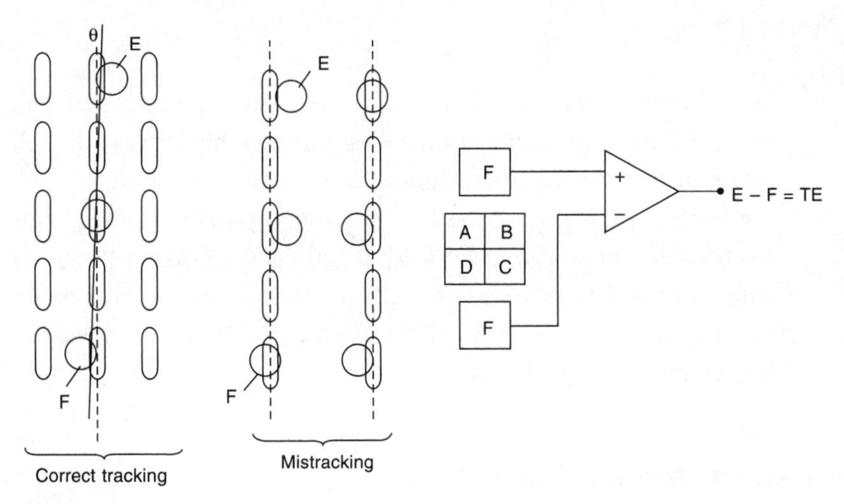

Figure 6.5

When the beam is on track, the outputs of diodes E and F are equal. However, if the beam leaves the track it will energize one of the diodes more than the other, producing a positive or negative tracking error. Different manufacturers use various techniques to correct this, all of which involve the use of additional photodiode assemblies.

The tracking and focus 2-axis device

The objective lens is contained in two separate coils, one for focus and the other for tracking control, placed at right angles to each other. The coils are placed inside a magnetic field produced by a permanent magnet, as shown in Figure 6.6. This assembly, known

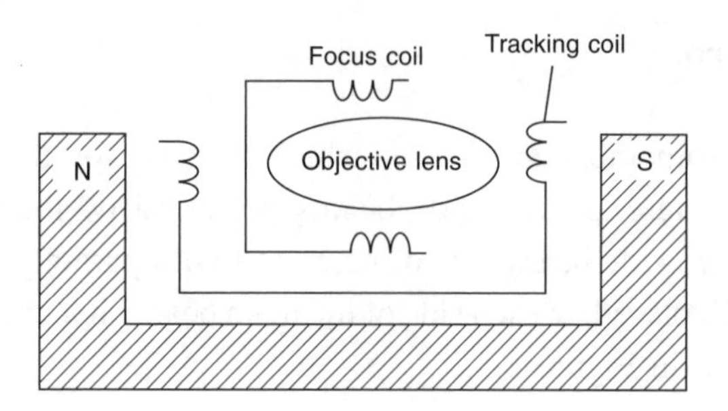

Figure 6.6

as a *2-axis actuator*, will move the lens in the horizontal and vertical directions. Current travelling through the focus coils will move the lens vertically to achieve focus, and current going through the track coil will move the lens horizontally (lateral movement) for tracking purposes. In DVD applications the 2-axis actuator is smaller than that used in CD applications because of the smaller pit size and the closeness of the tracks. In DVD, the device is known as a u2-axis device.

Playback compatibility

It is desirable for DVD-video players to play back traditional audio compact discs as well, a feature known as playback compatibility. The *optical pickup unit* (OPU) must be able to read both types of DVD discs (single-layer and dual-layer) as well as conventional audio CDs. This means that it must have the capability to generate laser beams of different wavelengths: a 650- or 635-nm red laser for DVDs, and a 780-nm blue laser for audio CDs. In addition, it must have the ability to focus at different depths to accommodate dual-layer DVD discs. This may be achieved in three different ways: by lens switching, bifocal lens, and liquid crystal shutter systems.

With the *lens switching* system the optical block contains two objective lenses, which are moved by a radial arm-type system. When playing back a DVD disc the DVD objective lens is used, and when playing back a CD the CD objective lens is used. These objective lens units will also have their respective tracking and focus coils included within them. Although this type of arrangement gives excellent playback it is cumbersome and expensive to produce, and has therefore only been used in a small number of machines.

With the *bifocal lens system*, only one objective lens is used. It combines both types of lenses required for CD and DVD playback into one unit. This is achieved by creating a hologram on the surface of the lens. The hologram consists of concentric grooves ranging from a few tens to a few hundreds of μm apart, resulting in deflection of the laser light at different angles. Depending on the type of disc inserted, one of the reflected beams will be used.

The *liquid crystal shutter system* changes the numerical aperture of the lens, thus changing the depth of the field to accommodate the two types of discs. The technique is similar to that used in an ordinary camera, which restrict the passage of the light beam through a shutter system. In the case of a DVD unit, the process is carried out by the use of a simple liquid crystal device.

CHAPTER 7

SIGNAL PROCESSING AND CONTROL

Introduction

Figure 7.1 shows the principle units involved in signal processing in a DVD playback system. The optical pickup unit (OPU) generates a laser beam with the appropriate wavelength and detects the reflected beam, using a number of photodiodes, to produce two types of signals: a varying signal representing pit and lands on the surface of the disc, known as the RF; and a number of error signals (not shown) used for the servo control system. All these signals are fed into the RF amplifier. The RF amplifier performs two distinct functions regarding the incoming signals; it amplifies the RF signal to a level suitable for processing by the next stage, and carries out digital processing of the error signals (not shown) for the servo DSP. The RF signal is fed to the RF processor, which carries out the following functions:

- RF signal processing, including demultiplexing to extract the various video, audio and other PES packets
- Clock sync extraction
- EFM+ (16/8) demodulation
- Frame/sector sync detection
- Error correction.

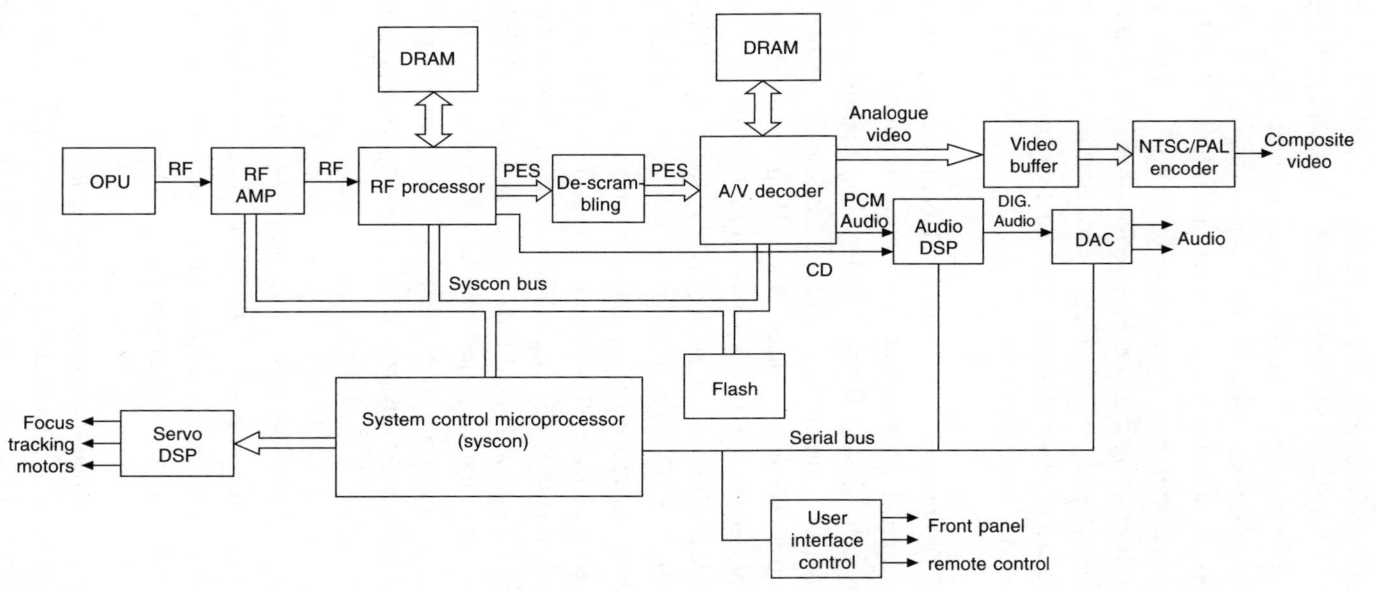
DRAM
DRAM
OPU
RF
RF
AMP
RF
RF processor
PES
De-scram-
bling
PES
A/V decoder
Analogue
video
Video
buffer
NTSC/PAL
encoder
Composite
video
PCM
Audio
Audio
DSP
DIG.
Audio
DAC
Audio
CD
Syscon bus
Flash
Focus
tracking
motors
Servo
DSP
System control microprocessor
(syscon)
Serial bus
User
interface
control
Front panel
remote control

Figure 7.1

The process of demultiplexing the PES packets requires a memory bank that stores the PESs as they arrive to allow the RF processor to select and rearrange them in the required order. The output from the RF processor is a series of video, audio and other PES packets relating to the selected piece. Before audio and video decoding, the PES stream has to be descrambled to return it to the original bit sequence. Following this the PES packets are sent to the audio/video (A/V) decoder chip, which carries out the decoding of the video, audio and sub-picture packets into digitized analogue video and pulse code modulated audio signals.

Decoding compressed video involves data decompression and the reconstruction of the video stream, picture by picture, from the I-frames and the P- and B-frames. Decoding therefore requires a memory bank to store the I-, P- and B-frames so that reconstruction of actual frames can take place. A similar process takes place in the case of decoding compressed audio information. The memory bank also serves as a 1-second audio delay for lip sync. This delay is necessary because the decoding of the video packets takes a longer time than the decoding of audio information. The A/V DRAM memory chip provides this memory bank. The video output is a digitized analogue Y, C_R, C_B multiplexed on an 8-bit parallel bus. The audio output is a serial pulse code modulated (PCM) audio stream, which is fed to an audio digital signal processor (DSP) for conversion into a digitized audio signal. This is then supplied to a digital-to-analogue converter to produce a two-channel stereo sound.

Before any further processing, the digitized video is fed into a video buffer. The purpose of the video buffer is to store video frames in order to ensure seamless video display in cases when the optical head is re-reading (or searching for) a sector on the disc.

As it stands, the video information cannot be displayed on a television receiver; it must first be encoded into a television system such as NTSC or PAL. This is carried out by the *NTSC/PAL encoder* (also known as *video encoder*), which produces NTSC- or PAL-formatted video signal for display on a TV receiver.

The whole process is programmed and controlled by a powerful microprocessor system that is centred on a *system control* (*sys con*) chip with an embedded RISC microprocessor core. The sys con carries all the necessary programming and controls for signal and servo processing. It incorporates data, address and control bus structures together with a FLASH memory store. It also provides a serial control bus for controlling such units as the audio DSP and ADC, as well as the user interface. The FLASH memory is a non-volatile RAM chip used to store start-up and other initializing and processing routines.

One of the functions of the sys con chip is to ensure that the correct track sector is read at the correct speed and with correct focus. This is carried out by the servo DSP chip.

The final part of the DVD player is the user interface, which is managed by the interface control. This provides the necessary interface between the player and front panel control buttons, front panel display and remote control handset.

Chip set

Increased integration has reduced the chip count of a DVD player, using the highly integrated system-on-a-chip (SoC) or the more recently system-in-package (SiP). Philips, for instance, uses an SoC incorporating the functions of the sys con and the A/V decoder, and Sanyo employs a single chip to integrate the functions of the RF processor and A/V decoder. A second-generation chip set has been developed, comprising fewer LSI chips, to cover even more functions.

The RF processor

The RF processor (Figure 7.2) has two major tasks; signal processing, and disc speed control. For this reason, this chip is also known as the RF processor/digital servo chip. These tasks must be carried out for both DVD and audio CD applications.

The input to the RF processor is a high frequency waveform known as the RF signal, which is generated by the optical head as it reads the pits and lands on the surface of the disc. The waveform is in analogue format and represents a digital bitstream. On a digital oscilloscope, which captures and stores part of the signal, the waveform is as shown in Figure 7.3. However, since it is not repetitive the waveform cannot be observed on an analogue oscilloscope, and the pattern shown in Figure 7.4 is displayed. This pattern, known as the RF *eye pattern*, is generated because the 8/16 (EFM+) modulation results in restriction of the number of ones or zeros that can follow in a sequence. The frequency of the pattern is determined by the closeness (i.e. the minimum size) of a pit. For DVD discs the pit size is small and hence the pattern has a higher frequency, requiring an oscilloscope with a minimum bandwidth of 40 MHz; in contrast, that produced by a CD disc only needs a bandwidth of 20 MHz.

Before processing can take place, the incoming analogue signal from the RF amplifier is converted into digital format by the ADC. The RF processor unit then produces a digital bitstream, which is used for signal processing and spindle speed control.

The bitstream from the RF processing unit contains 16-bit words, which are converted back into their original 8 bits by the EFM+ (8/16) demodulator. This is followed by error correction. It will be recalled that error correction bits are added to blocks of data, and therefore before error correction can be performed these blocks must be identified and reconstructed. This is done by the frame/sector sync detector. Once identified, the sectors are stored in the

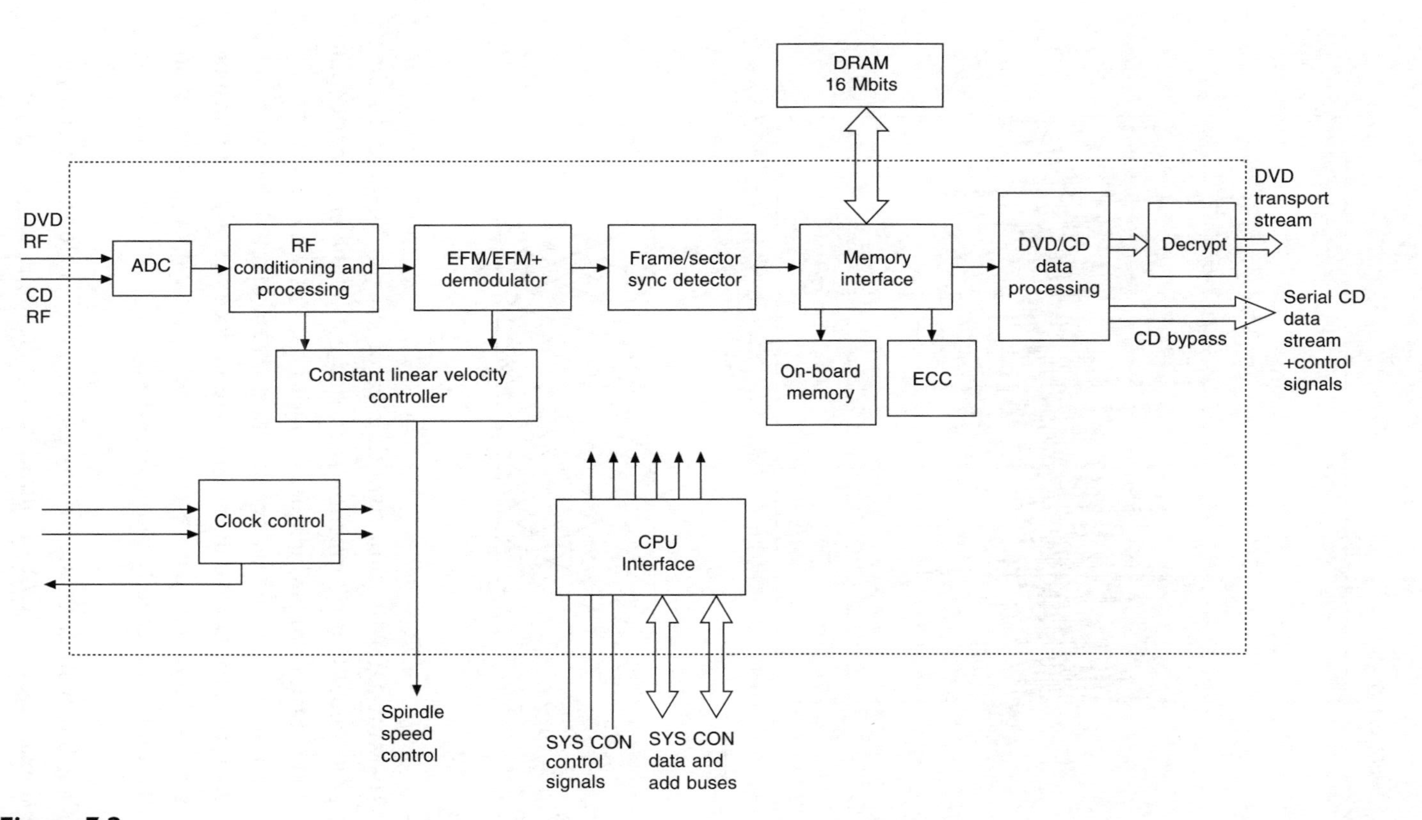

Figure 7.2
RF processor chip block diagram

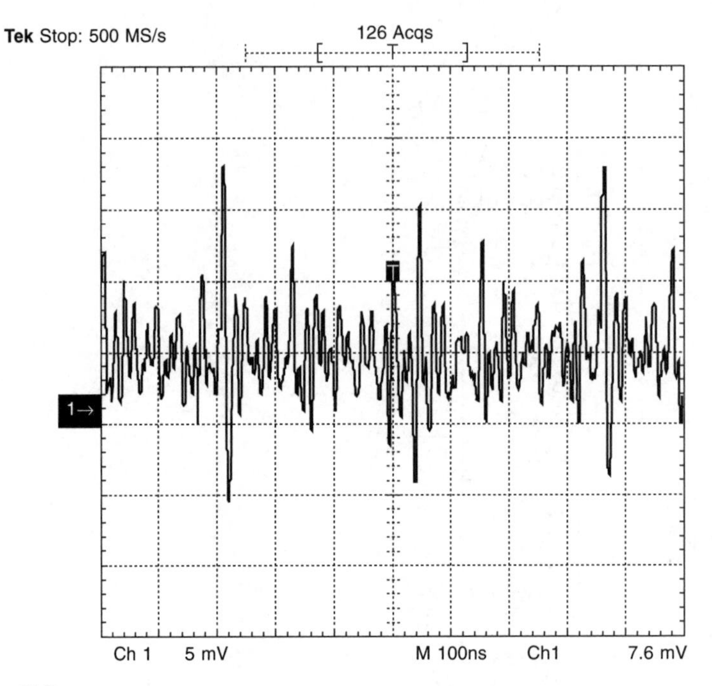

Figure 7.3

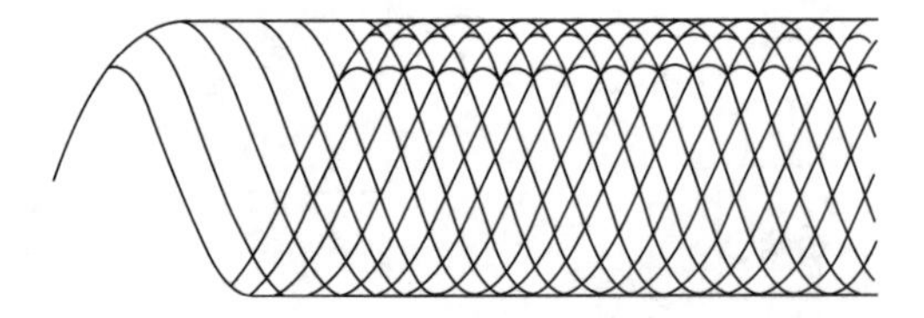

Figure 7.4

16-Mbit DRAM chip awaiting error correction and further processing. Error correction is carried out by the error correction code (*ECC*) unit, which uses an on-board as well as an external memory for the necessary data store. The corrected data stream is decrypted for DVD data stream to reproduce the original transport stream. Audio CD data streams do not require decrypting, and are therefore fed directly into the next stage.

The bitstream from the RF processor unit is also fed into a constant linear velocity (*CLV*) controller, which monitors the bit rate to determine the linear speed of the disc, and a signal is sent to the servo controller to change the speed of the spindle motor as appropriate.

Servo control

Figure 7.5 illustrates the main components of the servo control system of a DVD player. The RF amplifier receives the RF signal, which it passes to the RF processor via a low pass filter (LPF). It also receives photodiode signals (A, B, C, D, and so on) from the optical head for both DVD and CD applications. The photodiode signals are processed by the digital servo to produce three error signals – focus error (*FE*), tracking error (*TE*) and pull-in (*PI*) – which are fed into the servo DSP for further processing. In addition some players employ a fourth error signal, namely *tilt* (or *skew*), to detect and compensate for warped DVD discs.

The servo DSP also receives a spindle control signal from the RF processor. The spindle control signal represents the data flow rate. A fast flow rate indicates a fast speed and *vice versa*.

The servo DSP processes the servo input signals and produces focus, tracking, spindle speed, sled and (depending on the manufacturer) tilt control signals, which are fed to the appropriate drivers on their way to the relevant actuators and motors. The servo DSP is fully programmed and controlled by the sys con microprocessor chip as shown.

Playback start-up routine

When the tray is closed, or when 'playback' is selected, a start-up routine is initiated. The purpose of the routine is to determine the size of the disc, whether it is a CD or a DVD disc (and, if the latter, whether the disc is single or a dual layer), and to carry out what is

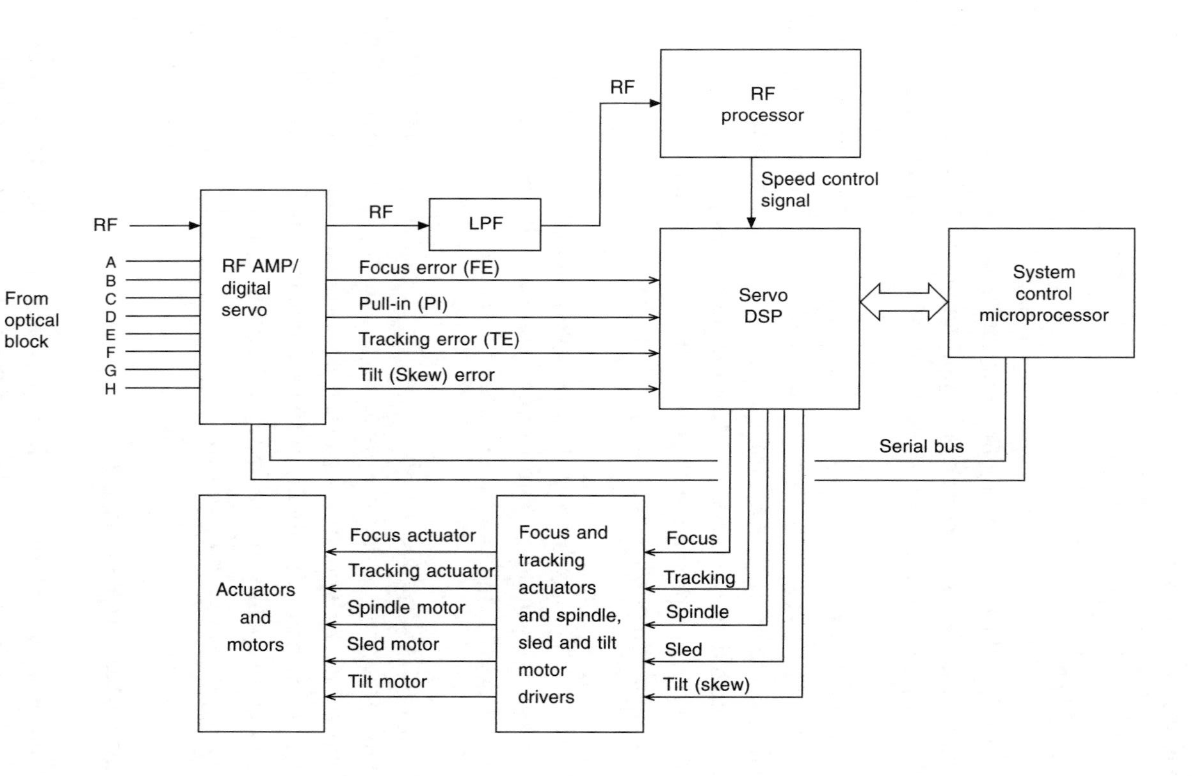

Figure 7.5
DVD servo control system

known as a focus search. The start-up routine will initiate the following steps:

1. The spindle motor rotates the disk at a relatively high speed
2. The laser beam is directed onto the disc and the reflected beam is detected by the photodiodes to produce the RF signal
3. The sled motor moves the optical head across the disc from the centre towards the circumference and back again to determine the diameter size and generate a tracking error signal
4. The two-axis actuator moves the objective lens up and down to obtain a focus, a process known as focus search.

CD/DVD detection

When the optical pickup head moves across the track, it generates a tracking error signal. If the tracking error voltage is low (0.4 V or less), then the inserted disk is a DVD. A CD disc, with its longer pits and wider track pitch, will produce a higher TE voltage of around 2 V.

Focus search

The focus servo loop is kept open while the optical head is searching to establish full focus. The objective lens is moved up and down along an S-curve crossing the full focus point, as shown in Figure 7.6. The RF signal strength increases as the beam goes towards the focus. When it crosses a specified threshold, determined by whether the disc is a DVD (SL or DL) or CD type, a Focus OK (FOK) goes HIGH and remains HIGH while the focus crosses the full focus (zero) line. The RF signal strength simultaneously goes to its maximum value and begins to fall again. At this moment, the servo loop is turned on and a focus is established.

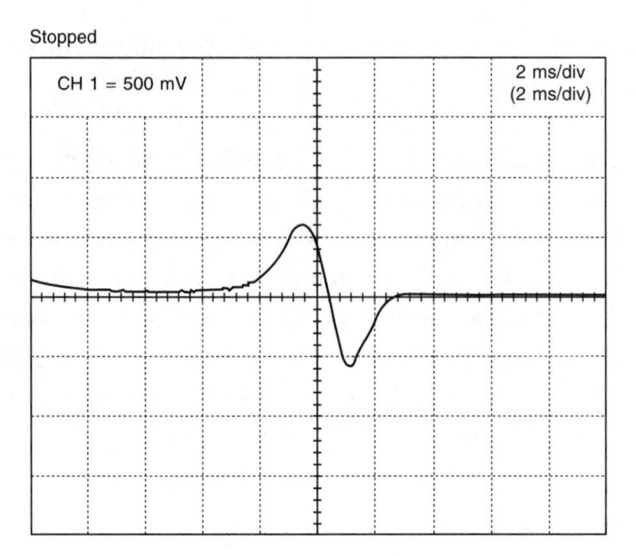

Figure 7.6
Focus search S curve

Single/dual layer detection

The difference between a single- and a dual-layer DVD disc is the intensity of the reflected laser beam. This is identified by what is known as the *pull-in* (*PI*) signal. The PI signal is obtained during the process of the focus search. A high PI signal of around 1 V indicates a highly reflective surface, and thus a single-layer disc; alternatively, a low PI of around 0.5 V indicates a dual-layer DVD disc.

Tracking control

Tracking is ensured by the horizontal actuator of the two-axis device for small adjustments, and by the *sled motor* for larger adjustments. The sled motor keeps the optical head moving along with the spiralling track and if necessary introduces a 'jump' when a new sector is selected for reading. The *sled error* (*SE*) is calculated inside the servo DSP, which monitors the *tracking error* (*TE*) signal. A typical TE signal is shown in Figure 7.7. The low frequency component of the TE signal is caused by the gradual spiralling of

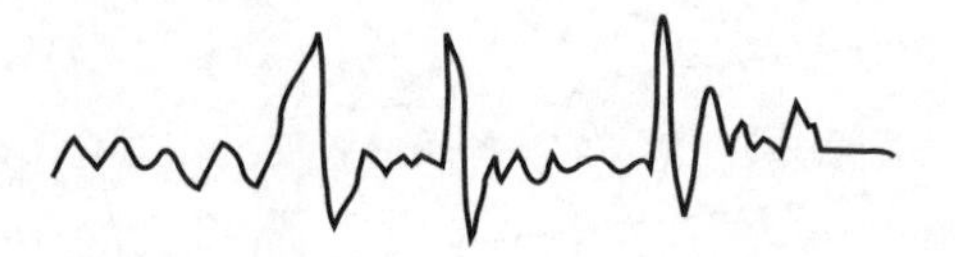

Figure 7.7
TE signal

the track towards the outside circumference, and is used to control the sled motor. The high frequency component, on the other hand, represented by the spikes, is caused by the jerky track itself. It is used to control the tracking actuator.

Tilt or skew control

Some types of DVD players include a tilt or skew control. Unlike CDs, DVD discs are built in two parts, each of which has a thickness of 0.6 mm. The two parts are then glued together to form a disc with a thickness of 1.2 mm – twice that of an audio CD disc. This construction is meant to prevent disc bend, which is caused by temperature and humidity changes. On the other hand, the thickness of the disc itself makes the reading of the disc more sensitive to very small warps in the disc surface. A warped disc will not present

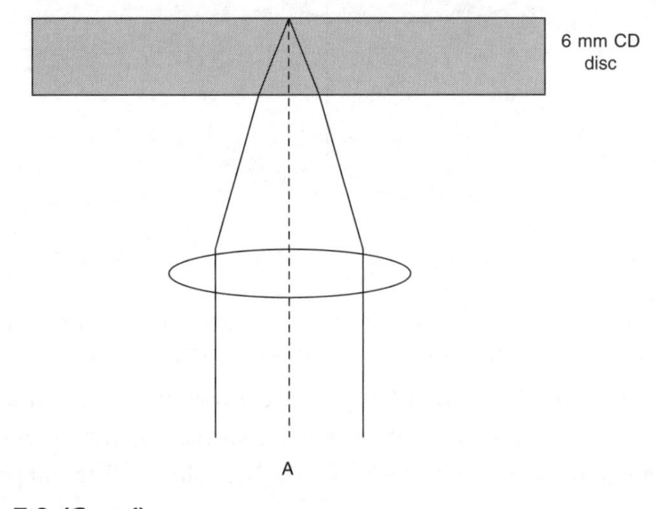

Figure 7.8 ***(Contd)***

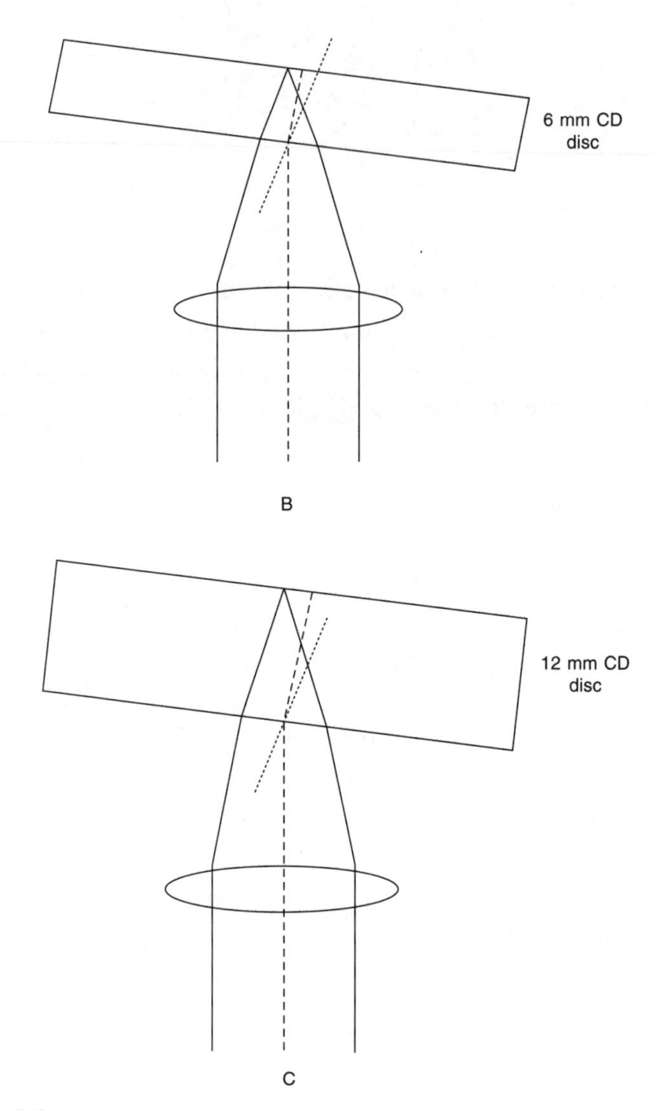

Figure 7.8

its surface at 90° to the beam, but will constantly change the angle of incidence as the disc rotates. When the disc is at right angles to the beam, the beam focuses on correct position on the track (see Figure 7.8a). However, when the disc surface is warped and is no longer at right angles to the beam, the beam hits a different position (see Figure 7.8b).

A changing tilt angle will cause what is known as a *skew error*, which will disturb the readout of the disc signal. The skew error depends on the tilt angle itself as well as the thickness of the disc. A thicker disc would cause a greater error even if the warp angle were very small (see Figure 7.8c). To detect this tilt, a skew sensor is mounted on the surface of the OPU. The skew sensor is built with two photodiodes (PDS1 and PDS2) and one infrared LED. If the disc is parallel to the OPU, the infrared light from the LED is reflected back equally on each photodiode. If the disc is warped and the disc surface is not parallel to the OPU, the intensity of the reflected light on the photodiodes will be different, resulting in a skew error signal. To avoid heavy compensations, an average of the skew error signal is used; this is the ratio of the difference to the sum of the photodiodes outputs:

$$\text{Skew error signal} = (\text{PDS1} - \text{PDS2})/(\text{PDS1} + \text{PDS2})$$

The error signal is sent to the control servo and the tilt motor is subsequently activated to make a correction that ensures that the OPU lens moves in parallel to the disc surface.

Practical RF amplifier/digital servo chip

Figure 7.9 shows a practical RF amplifier/digital servo chip showing the main test-point waveforms.

RF input from the optical head is fed into pin 1. The signal observed on an analogue oscilloscope is the 'eye' pattern described earlier. While the pattern will be the same regardless of whether a DVD or an audio CD is being played, the frequency of a DVD playback eye pattern is higher than that of a CD playback. An RF output is produced at pin 54, and this is fed to the RF processor. Signals A–D from the optical head photodiodes are fed into pins 9–12 and pins 5–8. The other photodiode signals, E–H, are fed into pins 13–16. These signals are processed by the digital servo part of the chip

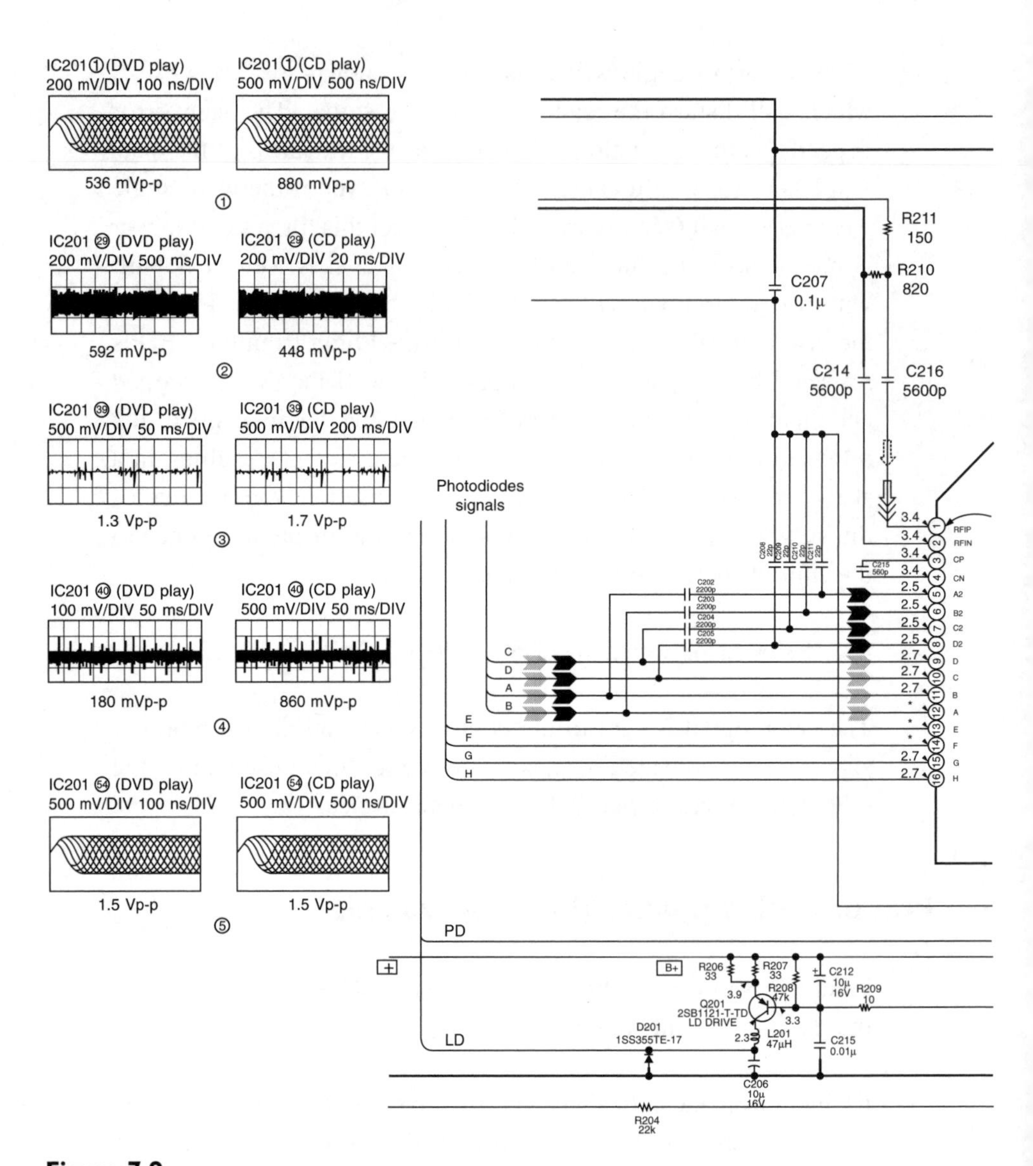
IC201 ① (DVD play)
200 mV/DIV 100 ns/DIV
536 mVp-p
IC201 ① (CD play)
500 mV/DIV 500 ns/DIV
880 mVp-p
①
IC201 ㉙ (DVD play)
200 mV/DIV 500 ms/DIV
592 mVp-p
IC201 ㉙ (CD play)
200 mV/DIV 20 ms/DIV
448 mVp-p
②
IC201 ㊴ (DVD play)
500 mV/DIV 50 ms/DIV
1.3 Vp-p
IC201 ㊴ (CD play)
500 mV/DIV 200 ms/DIV
1.7 Vp-p
③
IC201 ㊵ (DVD play)
100 mV/DIV 50 ms/DIV
180 mVp-p
IC201 ㊵ (CD play)
500 mV/DIV 50 ms/DIV
860 mVp-p
④
IC201 54 (DVD play)
500 mV/DIV 100 ns/DIV
1.5 Vp-p
IC201 54 (CD play)
500 mV/DIV 500 ns/DIV
1.5 Vp-p
⑤
R211
150
R210
820
C207
0.1μ
C214
5600p
C216
5600p
Photodiodes
signals
RFIP
RFIN
CP
CN
A2
B2
C2
D2
D
C
B
A
E
F
G
H
PD
LD
B+
R206
33
R207
33
R208
47k
C212
10μ
16V
R209
10
Q201
2SB1121-T-TD
LD DRIVE
D201
1SS355TE-17
L201
47μH
C215
0.01μ
C206
10μ
16V
R204
22k

Figure 7.9

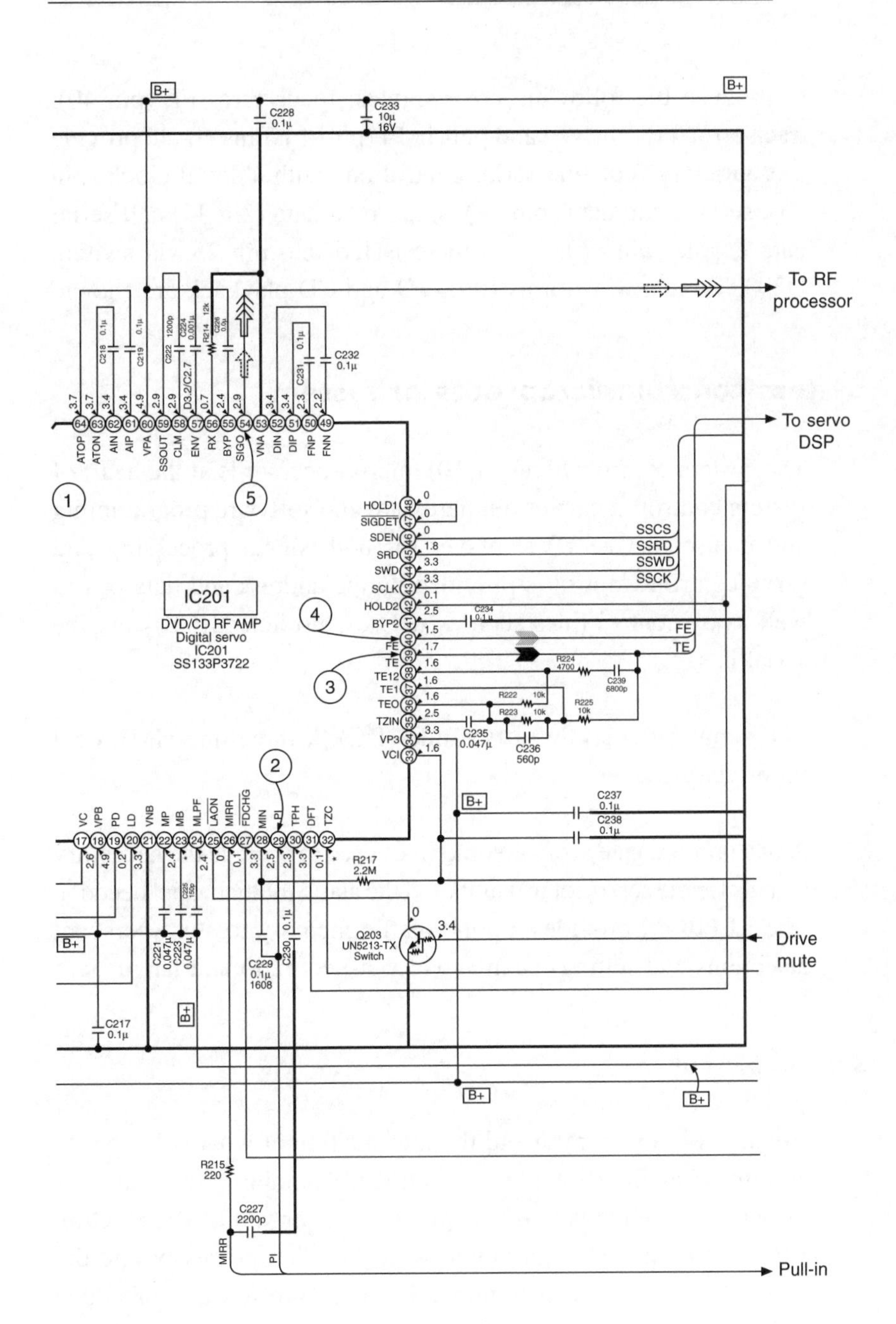

To RF processor
To servo DSP
IC201
DVD/CD RF AMP
Digital servo
IC201
SS133P3722
SSCS
SSRD
SSWD
SSCK
FE
TE
Q203
UN5213-TX
Switch
Drive mute
Pull-in
MIRR
PI

to produce the following error signals: focus error FE (pin 40), track error TE (pin 39), and pull-in PI (pin 29). Pins 43–46 provide a proprietary four-line serial control bus with a serial clock (pin 43), serial write data (pin 44), serial read data (pin 45) and serial data enable (pin 46). Drive mute is fed into pin 25 via switch, Q203. Typical waveforms for DVD and CD playback are shown.

System control microprocessor system

The system control (Figure 7.10) microprocessor is at the heart of system control. It carries out hardware and software programming and control of the DVD player for both signal processing and servo control. Control is provided by the address and data bus as well as by control lines such as IRQs, Chip Select (CS), and the serial bus.

The serial bus may be a two-line (I^2C), a three-line (IM), or a proprietary bus.

A one or more gate array is sometimes used to provide the necessary signals for the servo control units and the audio multichannel decoder. The EEPROM provides a non-volatile memory to store personal selections and settings such as video aspect ratio and language.

Sys con pin-out

Advances in integration and the introduction of what is known as system-on-a-chip (SoC) have resulted in combining the functions of two or more chips into a single IC. The pin-out of such a chip, which here integrates the function of the microprocessor and the gate array, is shown in Figure 7.11. The chip has an embedded RISC processor core for fast operation. It combines the functions of the microprocessor with that of the gate arrays, and feeds directly to the servo DSP. The chip has its own processing clock generated by a crystal, typically 12 MHz. The clock PLL provides the sys

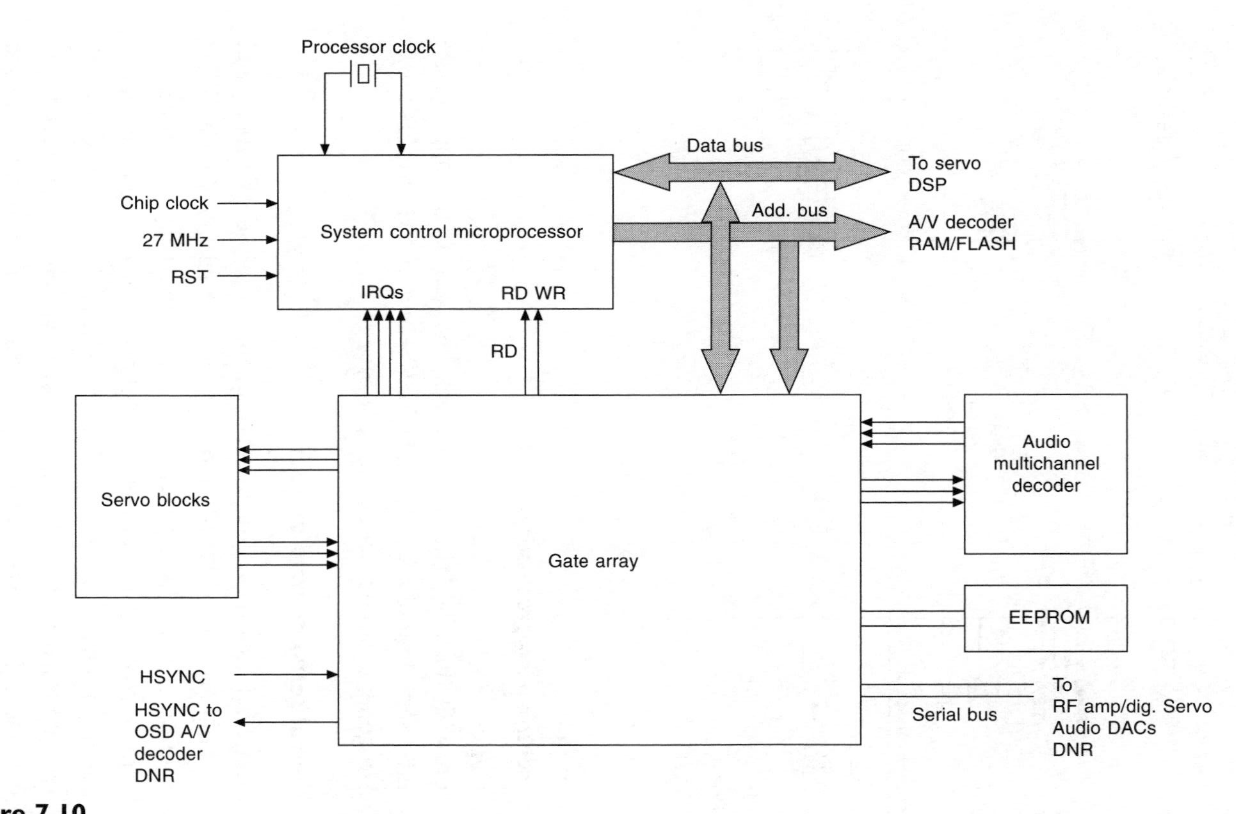

Figure 7.10
Sys con microprocessor system

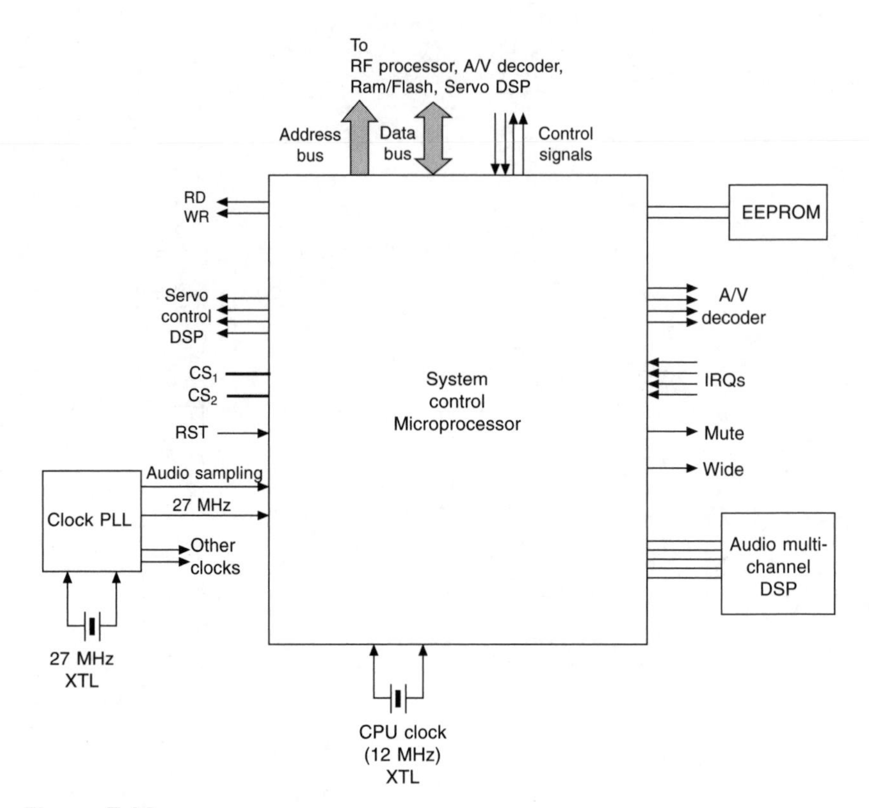

Figure 7.11
System control microprocessor pin out

con with the 27-MHz system clock, as well as the appropriate audio sampling clock (44.1 kHz for CD and 48 kHz for DVD). It also generates other clocks for the system.

Typical servo control pin-out

Figure 7.12 shows a typical servo control pin-out. The focus and tracking error signals (FE and TE) arrive from the RF amplifier/ digital servo at pins, 22 and 23, respectively. They are processed by the servo DSP to produce a pair of positive and negative control signals at pins 1, 3 and 5, 7. These are then fed to the focus and track coil drivers to position the objective lens for correct focus and track. The system provides for three motor servo controls: loading, sled and spindle. The loading motor drive is controlled by

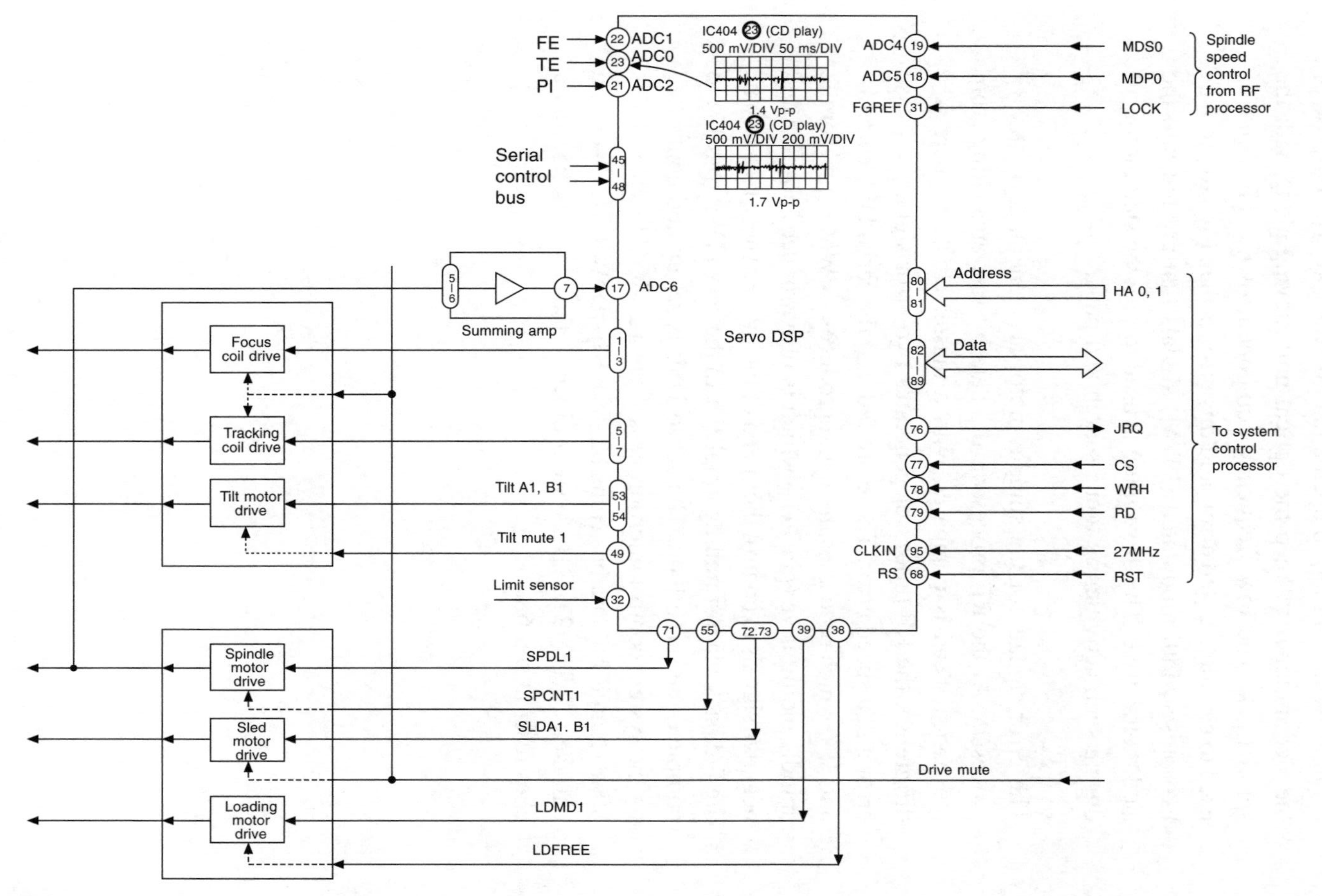

FE
TE
PI
ADC1
ADC0
ADC2
IC404 23 (CD play)
500 mV/DIV 50 ms/DIV
1.4 Vp-p
IC404 23 (CD play)
500 mV/DIV 200 mV/DIV
1.7 Vp-p
ADC4
ADC5
FGREF
MDS0
MDP0
LOCK
Spindle speed control from RF processor
Serial control bus
Summing amp
ADC6
Servo DSP
Address
HA 0, 1
Data
JRQ
CS
WRH
RD
CLKIN
RS
27MHz
RST
To system control processor
Focus coil drive
Tracking coil drive
Tilt motor drive
Tilt A1, B1
Tilt mute 1
Limit sensor
Spindle motor drive
Sled motor drive
Loading motor drive
SPDL1
SPCNT1
SLDA1. B1
Drive mute
LDMD1
LDFREE

Figure 7.12

a single signal (pin 39) to open and close the tray. The purpose of the sled motor is to keep the optical unit moving along with the spiral track. The low frequency component of the TE signal is used to calculate a sled error, and this error is then used to produce two pulse-width modulated (PWM) signals that are fed into the sled motor drive. The drive mute is used to turn the sled motor off during such activities as focus search and pause.

The error signals for the spindle motor are generated by the RF processor. At the RF processor, the data bit rate emanating from the optical head is compared with a master clock. If there is a difference, the RF processor generates two error signals: a phase error and a speed error. These are fed into the servo DSP (pins 18 and 19), which processes them and produces a PWM signal for the spindle motor drive (pin 71) – which in turn produces positive and negative signals to control the speed of the motor. These two signals are fed back via a summing amplifier to the servo DSP (pin 17) to complete the servo loop. The servo DSP is set up and controlled by the system control microprocessor via a 2-bit address bus (pins 80, 81), an 8-bit data bus (pins 82–89), READ, WRITE, IRQ, and chip select. The 27-MHz system clock is fed in at pin 95, and a chip reset at pin 68.

CHAPTER 8

VIDEO AND AUDIO DECODING

Introduction

Figure 8.1 shows the basic elements of the video/audio decoding part of a DVD player. The waveforms shown are those for a colour bar display. Multiplexed PES packets from the RF processor are processed by the V/A decoder to reproduce the original digitized luminance Y and the two chrominance colour difference components C_R and C_B. They are multiplexed on an 8-bit bus as shown. These are fed into a digital PAL/NTSC encoder to produce analogue video signals, which may be used to picture reproduction on a television receiver. Three sets of video signals are produced by the PAL/NTSC encoder:

1. Luminance Y and chrominance C
2. Composite video
3. R, G, and B.

The first set is for the S-video output port. The other two sets, composite video and RGB, are available at the SCART connector. As far as picture quality is concerned, the difference between the three outputs is very small and is not discerned by many people. Nonetheless, an S-video connector provides the best quality video, RGB is second best, and composite video is third.

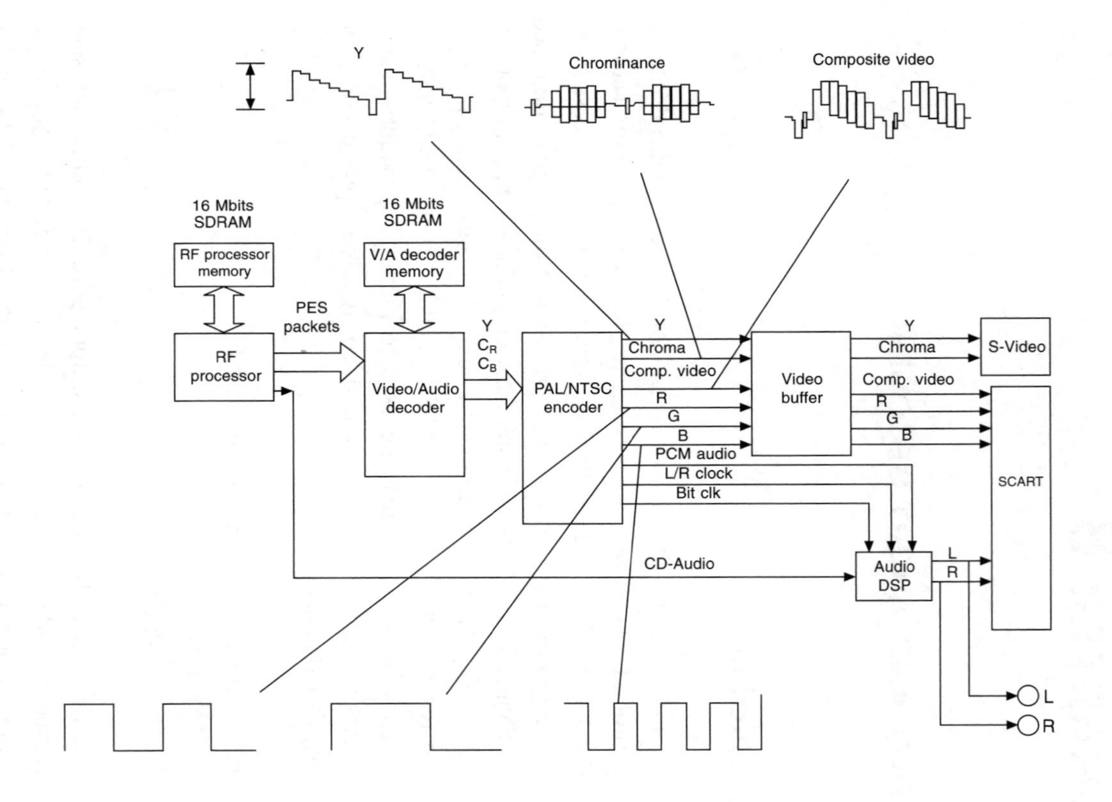

Y
Chrominance
Composite video
16 Mbits SDRAM
RF processor memory
16 Mbits SDRAM
V/A decoder memory
PES packets
RF processor
Video/Audio decoder
Y
C_R
C_B
PAL/NTSC encoder
Y
Chroma
Comp. video
R
G
B
PCM audio
L/R clock
Bit clk
Video buffer
Y
Chroma
Comp. video
R
G
B
S-Video
SCART
CD-Audio
Audio DSP
L
R
L
R

Figure 8.1

Before these video signals are fed into their appropriate output connectors, they are supplied to the video buffer to ensure a continuous video stream if the data flow from the optical head is interrupted by such things as searching for a sector, changing layers, etc.

DVD audio is decompressed and decoded into a pulse code modulated (PCM) data stream, which, together with control lines L/R, clock and bit clock, are fed into the audio DSP to produce L and R stereo sound that is fed to the SCART connector. They are also made available on individual phono connectors. CD audio output does not require decoding, as the data is recorded in PCM format without any compression. It is therefore fed directly into the audio DSP as shown.

The V/A decoder

The purpose of the video/audio decoder is to select the relevant video and audio packets, arrange them in the correct order, ensure synchronization between the video and audio (what is known as *lip sync*), and decompress them to restore them to their original form. Figure 8.2 shows a block diagram of a typical A/V decoding chip.

The incoming bitstream consists of multiplexed video, audio and other packetized elementary streams (PESs). The first task of the A/V decoder is to demultiplex the PES packets, extract the required packets, and store them in the SDRAM memory chip.

Video PESs are fed into the video decoder for data decompression and picture reconstruction, using the SDRAM memory to store picture frames as necessary. The output of the video decoder, which is in the form of multiplexed Y, C_R and C_B, is mixed with any onscreen display (OSD) generated by the processor before being made available on an 8-bit parallel bus.

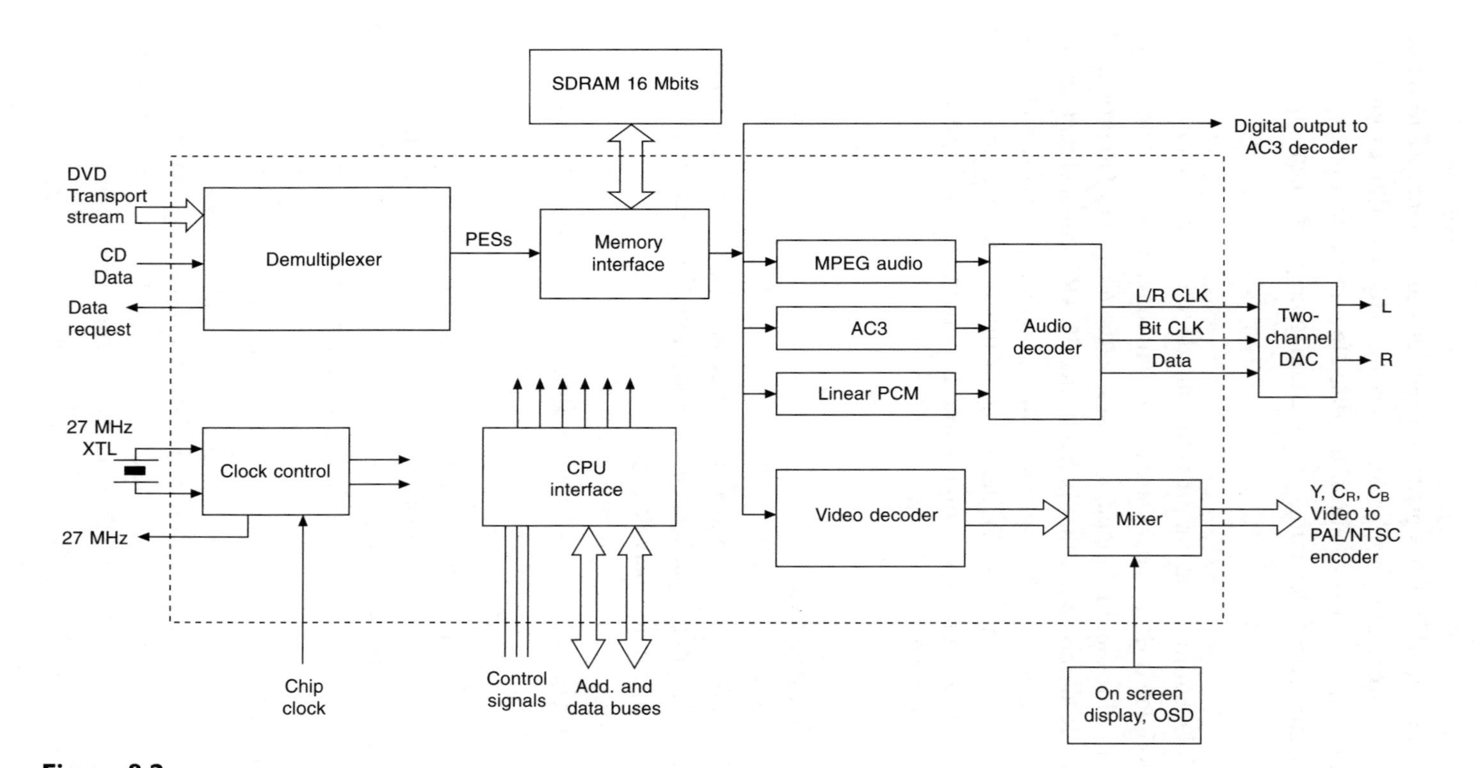

Figure 8.2
A/V decoder block diagram (version 2)

Audio PESs are fed into their appropriate PES decoder: MPEG, AC-3 or linear PCM. This is followed by the audio decoder, to reproduce L/R stereo or multi-channel outputs. The audio signal from the audio decoder is in the form of a data line carrying serial PCM together with the bit clock and L/R clock lines. Since video data require more time to be decoded than audio signals, a time delay is introduced in the audio path to ensure lip sync with the video information, frame by frame. The same video SDRAM memory chip is used for that purpose. Digital audio and video outputs are normally also available on coaxial or fibre-optic connector ports, or both. The 27 MHz system clock is provided by a crystal oscillator. The chip also requires its own chip clock, which is provided externally.

A modern V/A decoder chip incorporating the PAL/NTSC encoder is shown in Figure 8.3. PES packets enter the chip on an 8-bit bus on pins 31–38, and audio CD on pins 22–26. Six output pins are available, providing the following analogue signals:

Pin 69	Green multiplexed with luminance Y
Pin 63	Red multiplexed with colour difference B – Y or C_B
Pin 66	Blue multiplexed with colour difference R – Y or C_R
Pin 80	Composite video
Pin 74	Luminance Y
Pin 77	Chrominance C

The system 27-MHz clock is fed into pins 180 and 182, the chip clock (33 MHz) into pin 29, and audio frequency sampling 512 FS, which is derived from the system clock, into pin 12.

Audio processing

DVD players support three audio encoding techniques, MPEG-2, Dolby Digital (AC-3) and Linear PCM, providing mono, stereo and multi-channel audio outputs. Figure 8.4 shows a simple two-channel audio processing block diagram. The V/A decoder receives

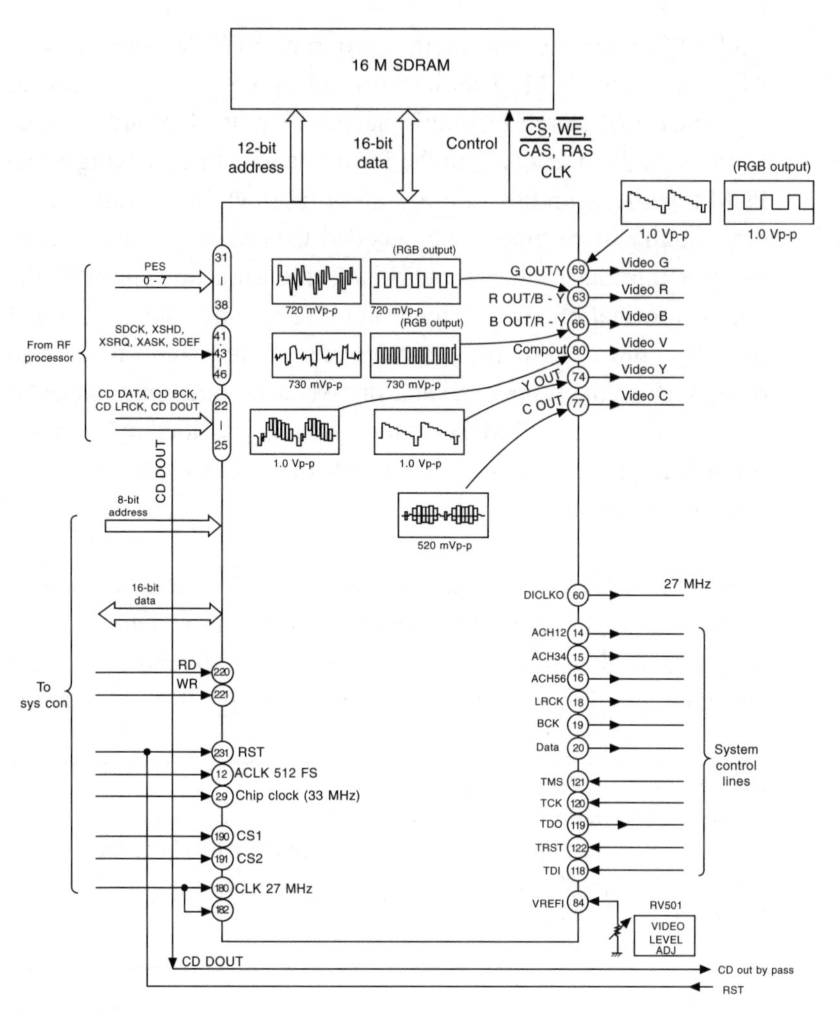

Figure 8.3

audio information in the form of multiplexed audio PES packets. The PES packets are demultiplexed and decoded to produce digitized serial audio data, which are fed into a two-channel digital-to-analogue converter to produce stereo sound.

The serial audio bitstream from the V/A decoder caries the codes of alternate samples of the L and R channels, as shown in Figure 8.5. Two control clocks are necessary to enable the audio data to be accurately converted into two-channel analogue output, namely

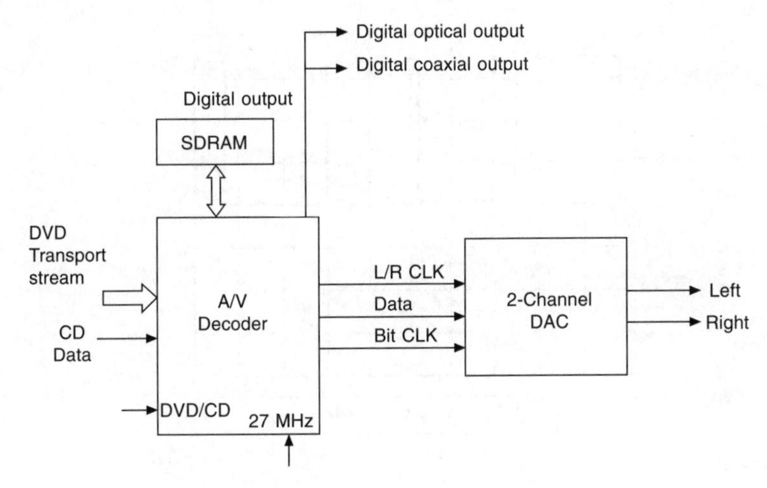

Figure 8.4

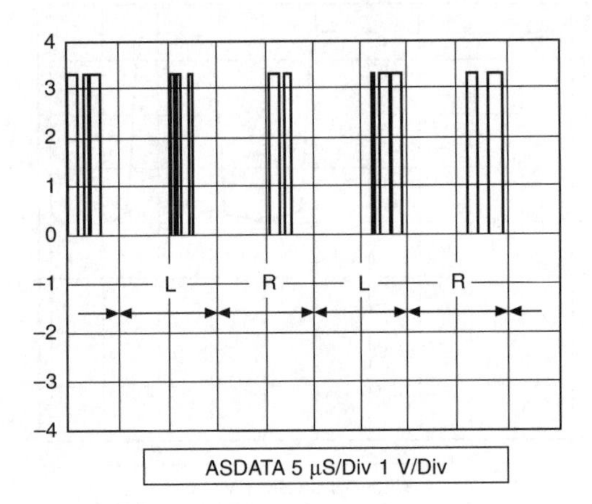

Figure 8.5

a L/R clock and a bit clock. The *L/R clock* (Figure 8.6) determines at any one time whether the bitstream represents the L or the R channel. The L/R clock has the same frequency as the audio sampling rate used, namely 48 kHz for DVD and 44.1 kHz for CD. The *bit clock*, which is in the region of 300 kHz (Figure 8.7), provides the clock for the digital-to-analogue converter.

The V/A decoder also provides a digital output on a coaxial or fibre-optic port for an external Dolby Digital decoder to produce multi-channel 5.1 surround sound.

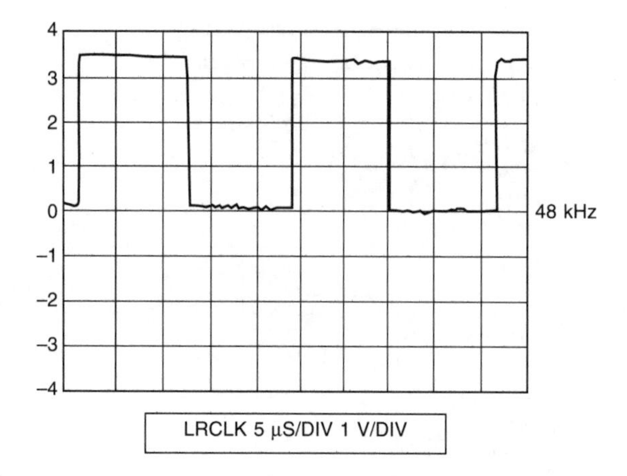

Figure 8.6

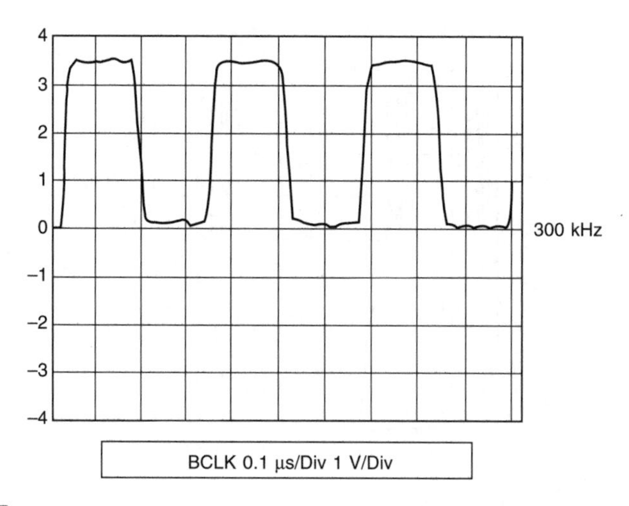

Figure 8.7

Multi-channel audio

Both MPEG-2 and Dolby Digital provide multi-channel audio outputs. Up to six channels are available in Dolby Digital, and up to seven channels are available for MPEG-2. In both cases, additional audio PES packets are required to be processed by the V/A decoder, which produces a number of serial audio data.

Figure 8.8 shows a Dolby Digital decoding and processing system.

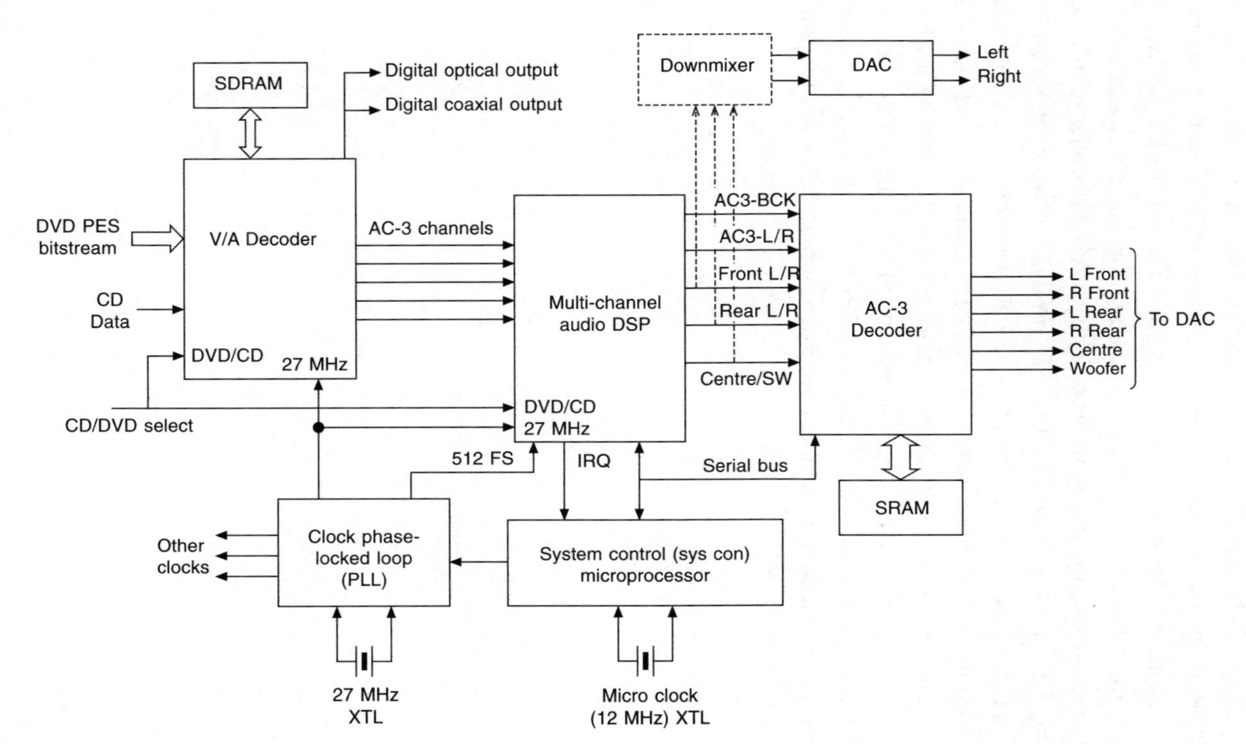
SDRAM
Digital optical output
Digital coaxial output
Downmixer
DAC
Left
Right
DVD PES bitstream
V/A Decoder
CD Data
DVD/CD
27 MHz
AC-3 channels
Multi-channel audio DSP
AC3-BCK
AC3-L/R
Front L/R
Rear L/R
Centre/SW
AC-3 Decoder
L Front
R Front
L Rear
R Rear
Centre
Woofer
To DAC
CD/DVD select
DVD/CD
27 MHz
512 FS
IRQ
Serial bus
SRAM
Other clocks
Clock phase-locked loop (PLL)
System control (sys con) microprocessor
27 MHz XTL
Micro clock (12 MHz) XTL

Figure 8.8

A phase-locked loop produces the 27-MHz system clock as well as other clocks including the 512 FS audio *block sampling* frequency. The multi-channel data lines from the V/A are fed into an audio DSP to produce six channels on three PCM serial data, two multiplexed channels on each data line: Front L/R, Rear L/R, and Centre/subwoofer, together with the appropriate clock and control lines. All six signals are then fed into an AC-3 decoder to produce the 5.1 Dolby Digital channels. To produce a two-channel output from a multi-channel stream, a *downmixer* has to be employed. Downmixing involves matrixing the centre and surround channels onto the main stereo channels. Downmixing will normally produce satisfactory results, but may be improved by tweaking the process for optimum results. In the case of MPEG-2 multi-channel audio, the centre surround channels are already matrixed onto the main stereo channels and thus do not require downmixing.

CHAPTER 9

POWER SUPPLY AND USER INTERFACE

Introduction

The operation of the DVD player – which includes play, pause, reject and standby – is carried out by the user either directly by pressing the appropriate button on the front panel, or indirectly by using a remote control handset. To facilitate this, an interface known as *user interface* is employed (see Figure 9.1). The power supply, which is of the switched mode type, provides a number of DC supplies for all units, including processors, decoders, drivers and motors. The user interface carries out the processing of the control signals from the front panel, and feeds them to the system control unit. Where a remote control is used, infrared signals have to be decoded before they are processed by the user interface. When the

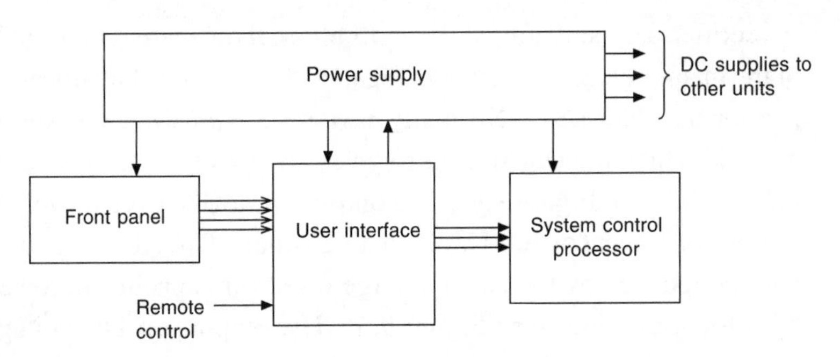

Figure 9.1

DVD player is in standby mode, the user interface remains alive, ready to respond to a button being pressed at the front panel or a signal from the remote control. When that happens, the user interface sends a signal to the power supply to revert it to the 'on' mode.

The power supply

The purpose of the power supply is to convert mains AC power into a stable DC supply of one or more voltages. A power supply consists of two parts: a rectifier and a regulator (Figure 9.2). The rectifier converts 50 Hz alternating voltage from the mains to a direct voltage. The purpose of the regulator is to ensure that the direct voltage from the rectifier remains unchanged, regardless of fluctuations in mains supply voltage or changes in the amount of current demanded by the load.

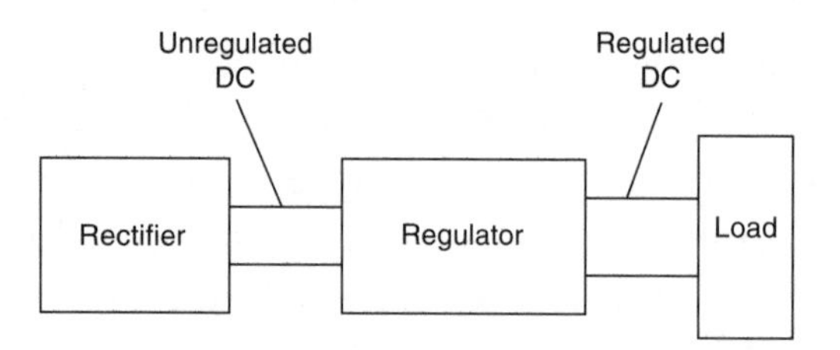

Figure 9.2

A rectifier can be a simple diode to block the negative half-cycle of the mains voltage, as shown in Figure 9.3. This type of rectification is known as *half-wave*. Normally, however, a *full-wave* rectifier is used, in which the negative half-cycles are converted into positive half-cycles (see Figure 9.4). The output has a DC level as well as a large component of 100-Hz alternating voltage. The AC component may be reduced by the use of a large reservoir capacitor followed by a low-pass filter (see Figure 9.5). The output is a DC voltage with a small alternating component, known as a *ripple*.

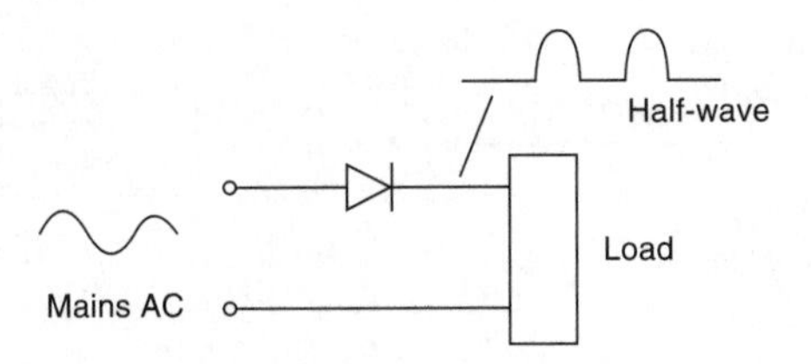

Figure 9.3

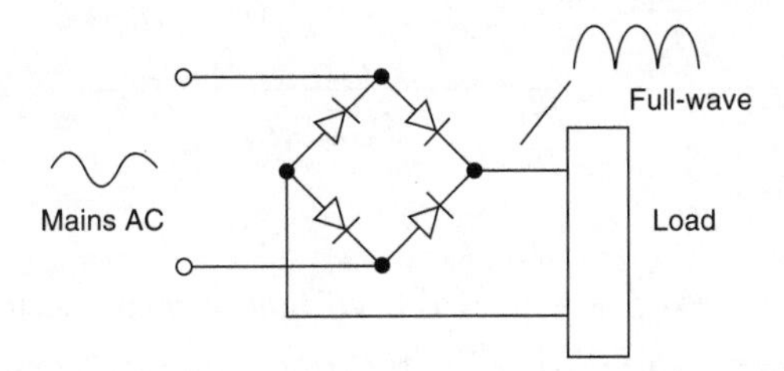

Figure 9.4

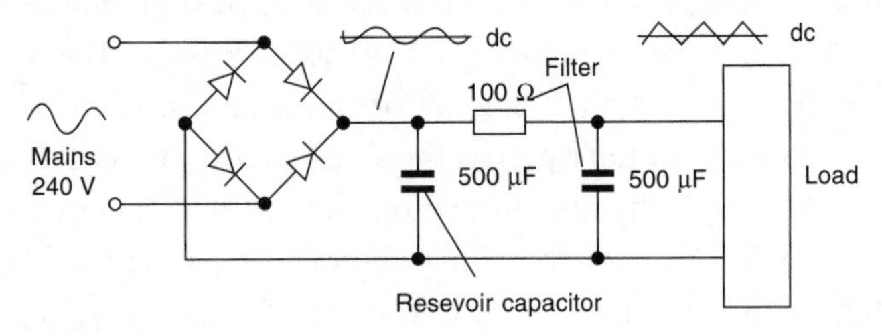

Figure 9.5

The regulator

In integrated circuits and most other applications, a highly regulated voltage supply is necessary. In the case of a microprocessor chip, for instance, the 5 V DC supply voltage V_{CC} must not vary by more than ±0.25 V. The function of the regulator is to ensure this. One of the most efficient regulators is the switched mode type.

In the switched mode regulator (Figure 9.6), a witching device is used to regulate the voltage output. When the switch is closed, voltage is applied to the load; when the switch is open, the voltage is removed. The average voltage across the load thus depends on the length of time the switch is kept closed compared with the time

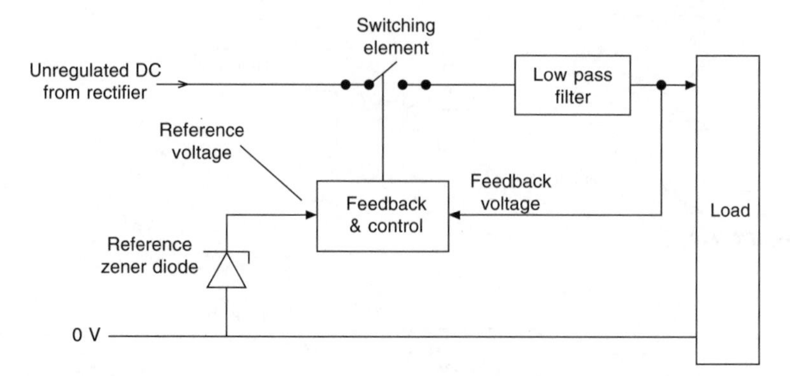

Figure 9.6

for which it is kept open. If the voltage across the load drops the switch is kept closed for a longer time, and *vice versa*. The timing of the switch is managed by a control unit, which senses the output voltage, compares it with a reference voltage, and sends a suitable pulse to turn the switch on and off as necessary. The switching action results in a small amount of ripple appearing at the output. The frequency of the ripple is the same as that of the rate at which the switching element is turned on and off, which is in the region of 3 kHz. Such a high frequency ripple makes filtering it far easier than it is for the ripple produced by the 50-Hz mains supply.

Control and regulation is only one requirement for a switched-mode power supply (SMPS). Of equal importance is the requirement for efficient use of energy, thus reducing heat loss and removing the need for a large heat sink, and resulting in a smaller physical size. Improved efficiency is achieved by the use of an inductor as an energy reservoir, as in Figure 9.7. When the switching element is closed, current I_1 flows from the positive side of the unregulated DC input into the load, keeping diode D_1 off. The magnetic field set up by the current flowing through inductor L_1 causes energy to be stored in the inductor. When the switch is opened, the current ceases and the magnetic field collapses. This causes a back e.m.f. to be induced across the inductor in such a polarity as to forward-bias D_1, causing current I_2 to flow into the load in the same direction as before. The energy stored in the inductor when the switch was

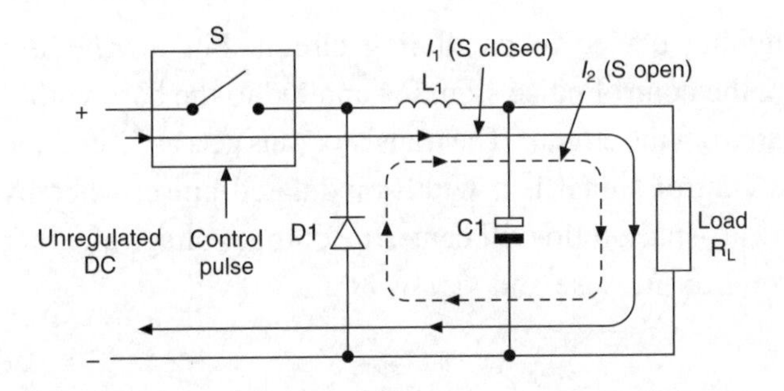

Figure 9.7
SMPS: series type

closed is thus utilized when the switch is open. The AC ripple at the output has a frequency of twice the switching rate, and is easily removed by a low-pass filter. The inductor normally doubles up as the primary winding of a high frequency transformer (see Figure 9.8). The transformer carries out two important functions; it provides a facility for stepping up or stepping down the voltage from the primary, and it also acts as an isolation transformer to separate the 'hot' earth of the mains side of the power supply from the 'cold' earth of the low-voltage DC side. The two earths must not be confused, especially when using test equipment.

The switching element may be a transistor, a thyristor, or any other

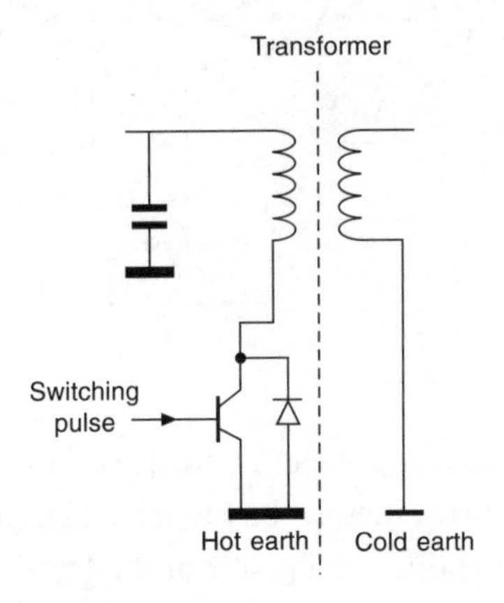

Figure 9.8

switching device or oscillating circuit. For a *transistor* (Figure 9.9), the control pulse signal is applied to the base with the output taken from the emitter. The transistor thus acts as an emitter follower. The control signal is a width-modulated pulse, whereby a larger mark-to-space ratio will cause the output voltage and therefore the current to increase and *vice versa*.

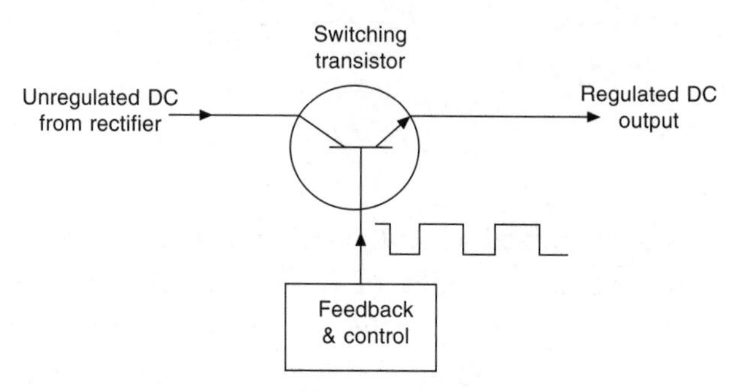

Figure 9.9

A *thyristor* has three electrodes; the anode, cathode and gate (see Figure 9.10). The control pulse is fed to the gate, which, when it goes positive, allows current to flow through to the load and *vice versa*.

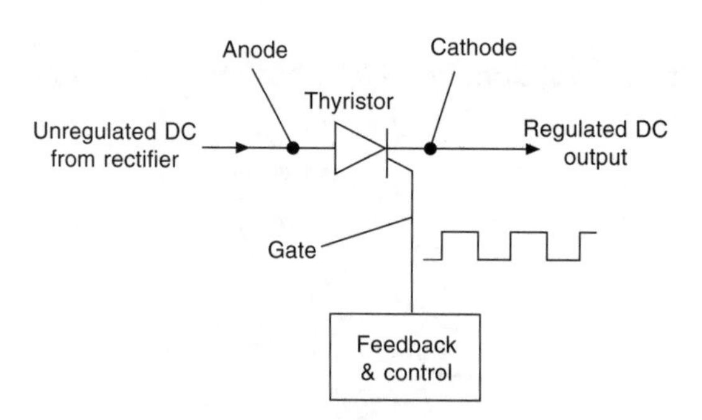

Figure 9.10

A more efficient arrangement is the use of a *self-oscillating circuit*, usually a *blocking* oscillator (see Figure 9.11). At switch on, transistor TR1 begins to conduct as a result of the forward bias applied to its base via the start-up resistor R1. The collector current increases,

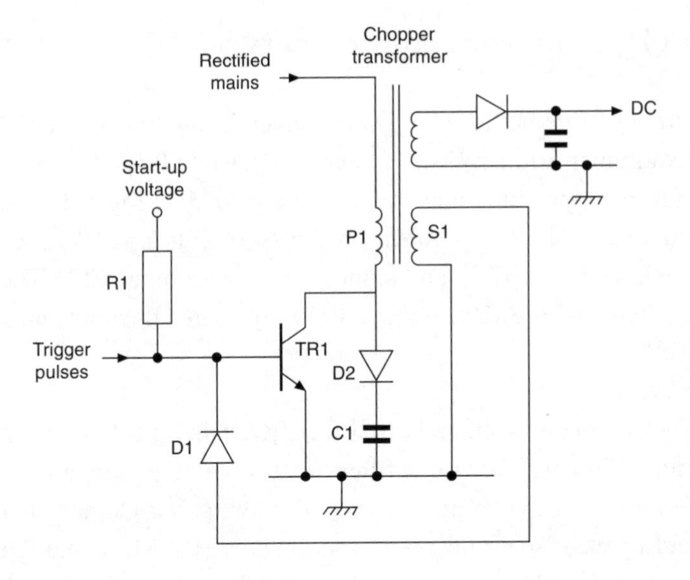

Figure 9.11
Use of a blocking oscillator

which induces a positive voltage across the secondary winding S1. This will forward-bias diode D1, and further increases the current through TR1. When saturation is reached, the increase in the current ceases and a negative voltage is induced across S1. This will reverse-bias D1, and switches off the transistor. At this point, the voltage across the primary winding P1 reverses and D2 switches on. The tuned circuit formed by primary windings P1 and C1 begins to oscillate. For the first half of the oscillation, energy is transferred from P1 to C1; for the second half of the oscillation cycle, energy begins to transfer back to P1 and D2 is reverse-biased. This will stop the oscillation, and the current through P1 reverses, causing a positive voltage to be induced across S1, turning TR1 on. The process then repeats itself. The frequency of oscillation is controlled by trigger pulses from the control unit of the SMPS. These pulses are fed into the base of the transistor to initiate each cycle, thus keeping the output constant.

DVD player power requirements

Different elements of a DVD player have different DC power requirements in terms of voltage and power. Generally, processing and memory chips may require 3.3 V or 5 V, logic devices 5 V, fluorescent indicator tubes on the front panel –10 V, focus and tracking coils –12 V, and spindle and sled motors 12 V. The main elements of a SMPS for a DVD player are illustrated in Figure 9.12.

The AC mains voltage is first rectified using a full-wave rectifier and, following a low-pass filter (not shown), the unregulated DC is fed into the switching element. The switching element is closed and opened by a pulse-width modulated (PWM) signal from the control unit. The switched DC output from the switching element is fed into the primary of transformer T_1. A number of secondary windings provide different output voltage levels. Each output is first rectified by a single diode, and is then fed through a low-pass filter to remove the high frequency ripple. The low-pass filter is a simple L/C π type with typical values shown in Figure 9.12. In the case of one or more DC outputs further stabilization may be necessary, as in the case of the 5 V supply line in the diagram. Negative voltages for the –10 V and the –12 V DC outputs are produced by simply reversing the diode. The feedback voltage is taken from one of the DC outputs (normally the 5 V or the 3.3 V line), via an opto-coupler, to the feedback and control unit. The opto-coupler consists of a photodiode and a phototransistor. The photodiode will emit light in accordance with the level of the output voltage, and the emitted light is sensed by the phototransistor, which causes it to conduct. The current through the phototransistor, which is determined by the strength of the emitted light from the photodiode, determines the mark-to-space ratio of the PWM waveform turning the switching element on and off. The use of a photocoupler ensures that the two earths (hot and cold) remain isolated from each other.

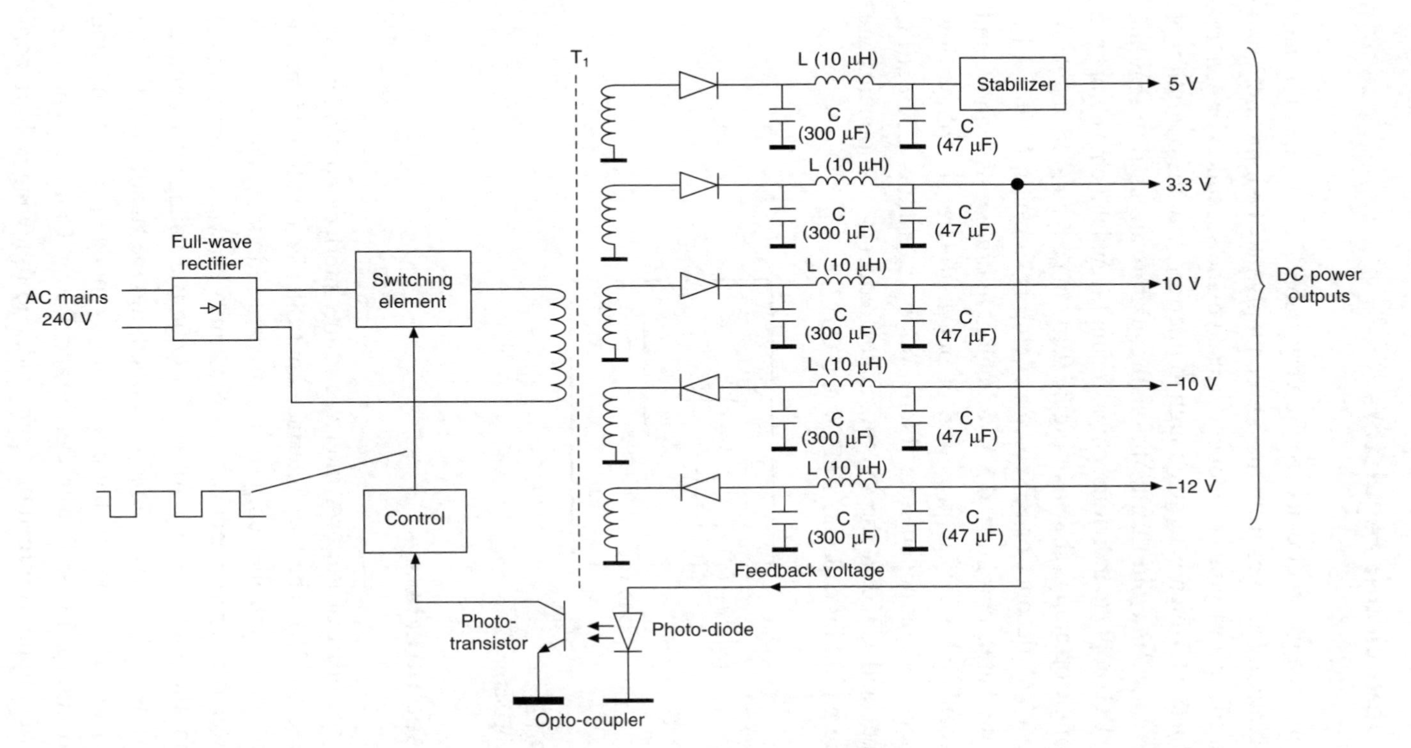

T1
L (10 μH)
Stabilizer
5 V
C
(300 μF)
C
(47 μF)
L (10 μH)
3.3 V
C
(300 μF)
C
(47 μF)
Full-wave
rectifier
Switching
element
L (10 μH)
10 V
DC power
outputs
AC mains
240 V
C
(300 μF)
C
(47 μF)
L (10 μH)
−10 V
C
(300 μF)
C
(47 μF)
L (10 μH)
−12 V
Control
C
(300 μF)
C
(47 μF)
Feedback voltage
Photo-
transistor
Photo-diode
Opto-coupler

Figure 9.12

Integrated circuit regulators

Some integrated circuits require very stable DC supplies for their operation. DC lines from the power supply may be long enough to introduce a varying component, caused by noise, spikes or general transients, which is large enough to interfere with the operation of the chip. To avoid this, additional regulators are used to regulate the DC supply to individual chips. A simple regulator chip composed a single op amp is shown in Figure 9.13, and a typical IC regulator for a V/A decoder or RF processor chip is shown in Figure 9.14. In the latter, unregulated 5 V voltage is fed into pins 1 and 5 and a regulated 3.3 V is produced at pin 4. Reference voltage low frequency components are filtered out by capacitors C319 at the input and C320 at the output. High frequency components are filtered out by capacitors C306 and C317.

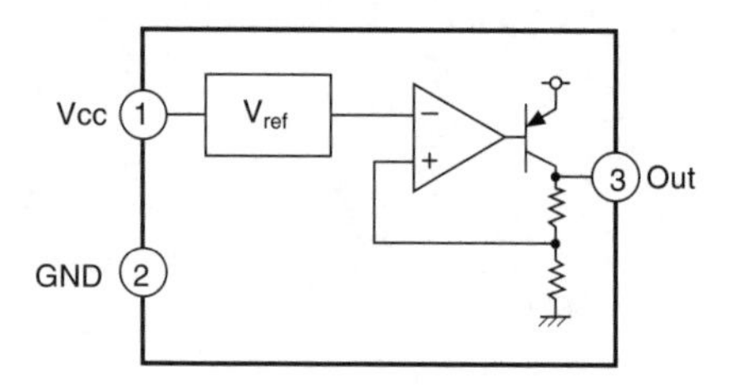

Figure 9.13
Voltage regulator

The user interface

At the heart of the user interface is the interface controller chip (Figure 9.15). The user interface controller is a dedicated micro-processor controller chip. It has its own individual clock, and is powered by a permanent voltage (sometimes known as 'ever' voltage) from the power supply. An *ever voltage* is a voltage that remains alive when the power supply is turned into the standby mode. The user interface controls the power supply mode of operation using two control lines: Power Detect and Power Control. When the power supply is turned on from cold, and following a short start-

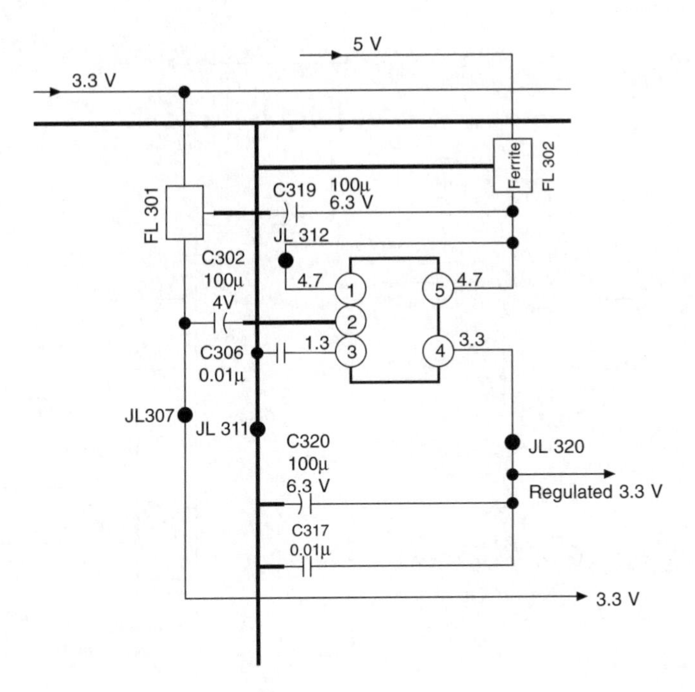

Figure 9.14

up routine, the power supply settles into the standby mode in which only the ever voltage (3.3 V in Figure 9.15) remains switched on. The player will remain in this mode until the button on the front panel or the remote control handset is pressed, or until the on/standby button is pressed. When this happens, and provided the power supply is sound (indicated by Power Detect line HIGH), the Power Control line from the interface goes HIGH, switching all the other voltages on, and the player goes into the 'on' mode.

The user interface has several other functions, including:

- Receiving and decoding signals from the remote control handset
- Receiving and decoding instructions from the front panel button switches
- Controlling the front panel display
- Setting the appropriate switching of the input and output ports to set the video and audio outputs, such as SCART and S-video sockets

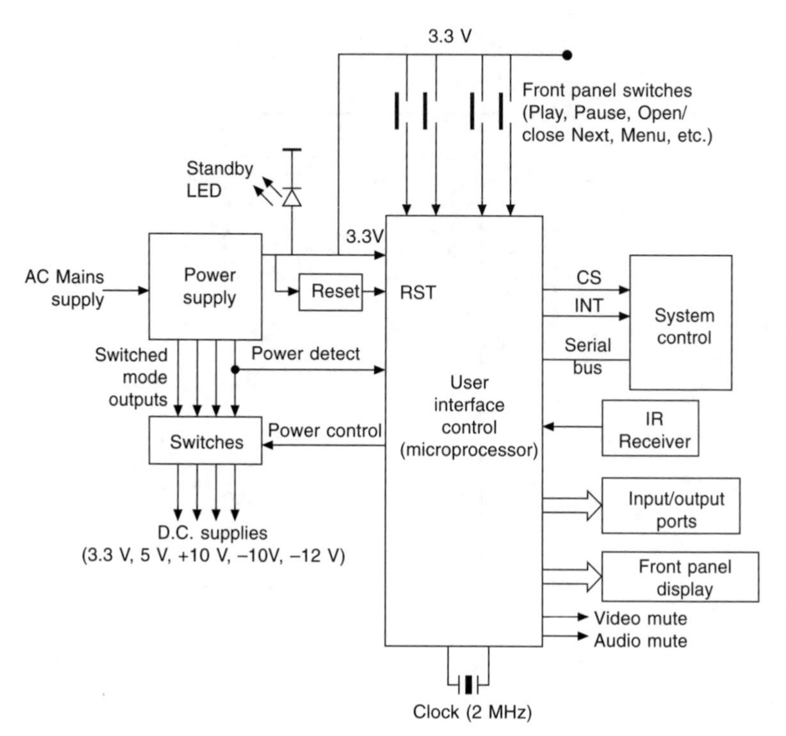

Figure 9.15
User interface controller

- Providing video and audio mutes as necessary
- Communicating with the system microprocessor controller via a serial bus and other control lines such as chip select (CS) and interrupt (INT).

A typical circuit diagram for a DVD player switched mode power supply is shown in Figure 9.16.

FET Q101, together with transformer winding W2, C117 and zener Q102, form a self-oscillating circuit running at a rate determined by the time constant R105/C117. The capacitor charges up through R105, and when it reaches a certain voltage it turns the transistor Q101 on, which discharges the capacitor, turning itself off in the process – and so on. Feedback control is provided by the network composed of the opto-coupler Pc101, transistor Q102 and zener

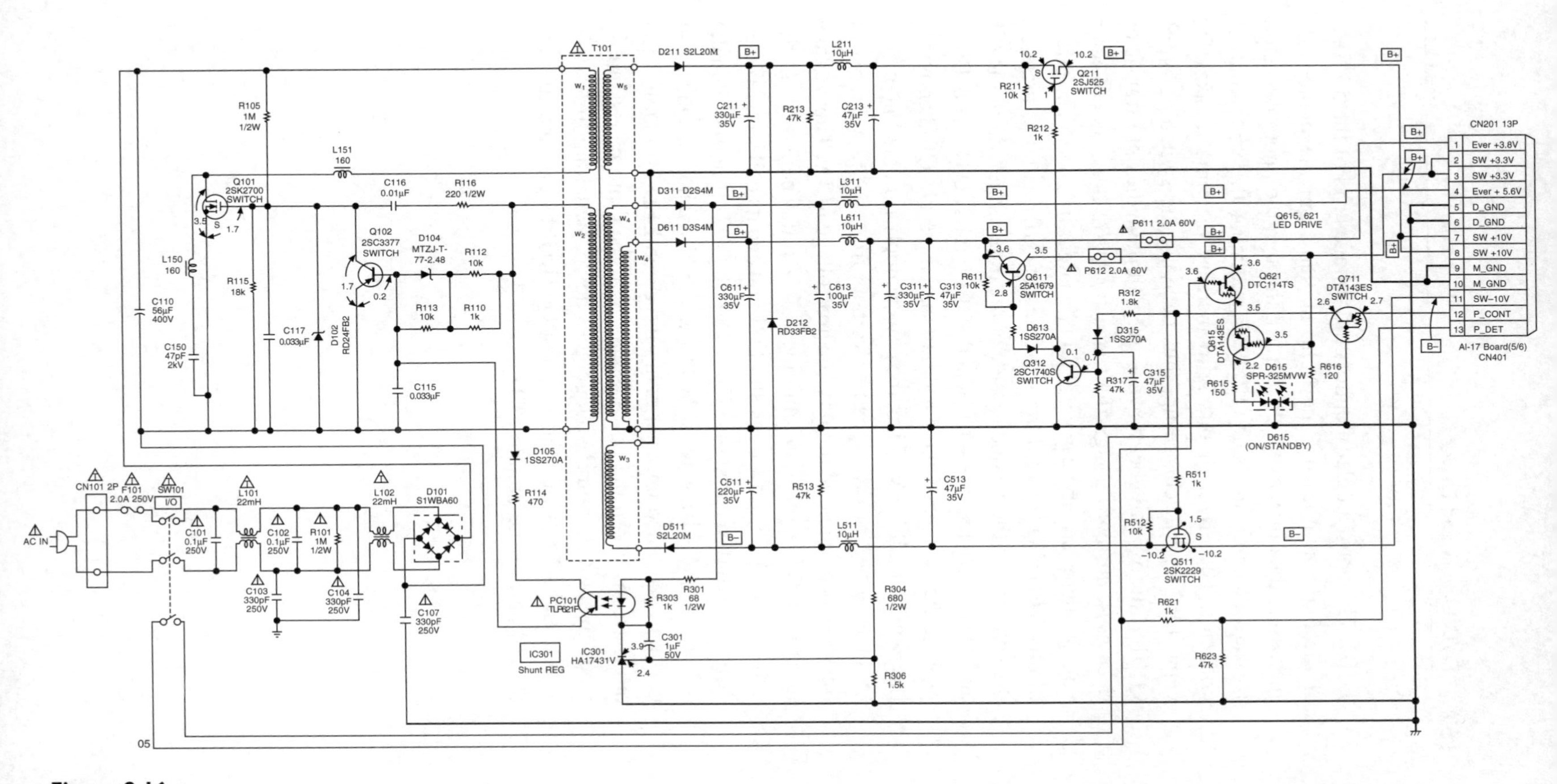

Figure 9.16
Power supply circuit diagram

D102. The opto-coupler monitors the voltage level of the 5 V rail. If the 5 V rail goes up, the light emitted from the photodiode increases, thus increasing the current through the phototransistor, forward-biasing transistor Q102 and reducing the voltage at its collector. The effect of this is to increase the time it takes for capacitor C117 to charge up, thus delaying the switching on of Q101. If the 5 V rail goes down, current through the opto-transistor decreases, current through the transistor Q102 falls, its collector voltage increases, and capacitor C117 charges up faster. The effect of this is to bring forward the switching of Q101 and thus increase the voltage. On the secondary side of the transformer, diodes D211 and D511 provide the +10 V and −10 V rails, D311 and D611 provide the ever 5.5 V and 3.8 V rails, and transistor Q611 provides the switched rail 3.3 V. When the power supply is in the 'on' mode, Q711 is turned on by a positive voltage from the power control line into its emitter, taking its collector voltage to 3 V. This voltage turns FET Q511 on, switching the −10 V rail. The 3 V from the collector of Q711 turns transistor Q312 on and its collector voltage drops, which turns transistor Q611 and FET Q211 on to switch the 3.3 V and the +10 V rails. With the 3.3 V available, Q621 and Q615 are switched on, energizing the green LED of double LED diode D615. When the power supply is turned into the 'standby' mode the power control voltage goes down, switching the −10 V, +10 V and +3.3 V rails off, turning Q621/Q615 off, and lighting the red part of the double LED diode.

Chapter 10

Servicing DVD players

Introduction

The initial purpose of the servicing of a faulty DVD player is to ascertain which of the major blocks of the player (Figure 10.1) suffers from malfunction.

RF and other analogue signals from the optical head are processed to produce video and audio signals for the SCART, S-video and other output ports. The processing section also sends focus, tracking and spindle speed error signals to the servo control as shown. Several memory stores are provided, including Flash memory for system control, where non-volatile (permanent) bitstreams such as start-up, video and audio decoding, and demultiplexing routines are stored; SDRAM for the V/A decoder; SRAM for the RF processor; and a video buffer, which ensures uninterrupted video display.

The user interface receives commands from the front panel or the remote control handset, decodes them, and communicates them to the system control microprocessor. The main clock is the 27-MHz system clock, which anchors the signal processing operation. Other clocks include the various chip clocks, which set the speed of operation of chips such as the system control, user interface and

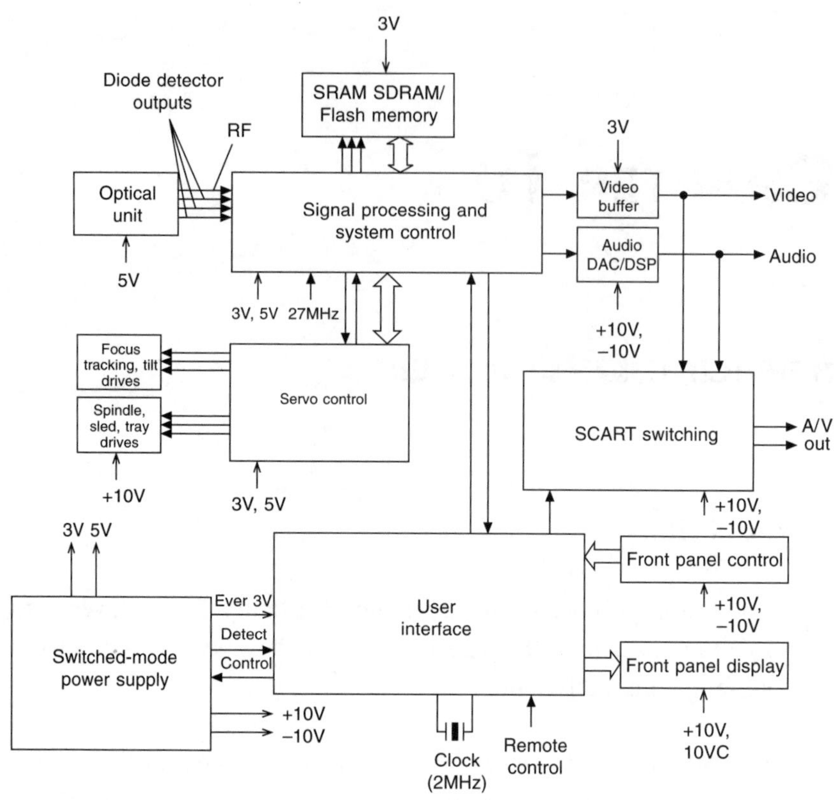

Figure 10.1
DVD servicing

RF processor chips. The power supply provides the necessary regulated DC voltage rails for all units.

Once a faulty block has been suspected, further tests must be carried out to ascertain the precise cause of the fault – such as a faulty component, logic chip or processing IC.

Being a microprocessor-based system, test signals are overwhelmingly a series of bits, which are undistinguishable from each other. The exceptions are clocks, Resets and DC power lines. Traditional test equipment, such as a DVM or an oscilloscope, is not suitable for testing a digital bitstream. For these purposes, logic-sensing instruments such as the logic probe and the digital storage oscilloscope (DSO) are necessary.

Logic-state test instruments

The logic probe (Figure 10.2) is a logic state test instrument that indicates the logic state at a test node. It can indicate logic 1, logic 0, an open circuit, and a data stream pulse. Two indicator LEDs are used to indicate a HIGH and a LOW. No light indicates an open circuit or an indeterminate logic state, and a flickering light or a special indicator indicates the presence of digital activity in the form of a bitstream. Use of a pulse stretcher, pulses as narrow as 10 ns may be detected. Although the logic probe cannot examine the actual waveform, it nonetheless provides a fast and simple method of testing for digital activity at various points along the signal path. Where real-time waveforms have to be examined, an oscilloscope has to be used.

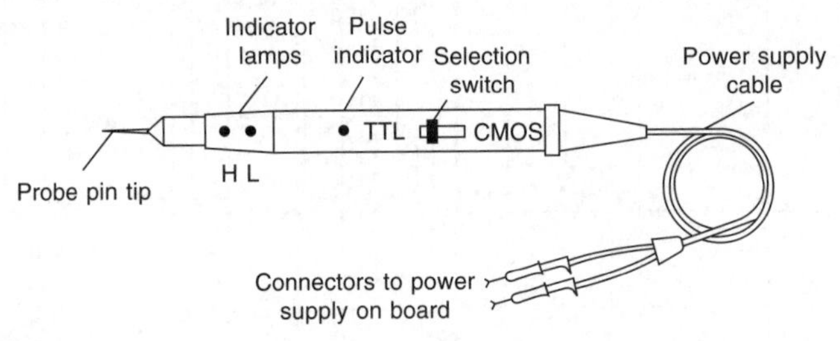

Figure 10.2
Logic probe

The cathode ray oscilloscope

The oscilloscope is used to display both analogue and digital waveforms, from which measurements of amplitude, frequency and time can then be made. Analogue signals in a DVD player are not dissimilar to those found on an analogue TV receiver, and may be displayed using a normal analogue oscilloscope. However, examining a data bitstream requires a digital storage oscilloscope with a minimum bandwidth of 100 MHz and a sampling rate of 500 million samples per second. Such an oscilloscope may be also used to display and examine analogue signals.

The storage oscilloscope captures a part of the data bitstream, stores it, and then displays the waveform on the screen for examination and measurements. This may then be repeated for another part of the data stream, and so on. Figure 10.3 shows a typical data stream captured by a digital storage oscilloscope. Unlike analogue systems, where a test point has its own unique signal in terms of waveshape, frequency and amplitude, a sequence of ones and zeros in a data stream has the same general wave shape and amplitude regardless of the test point.

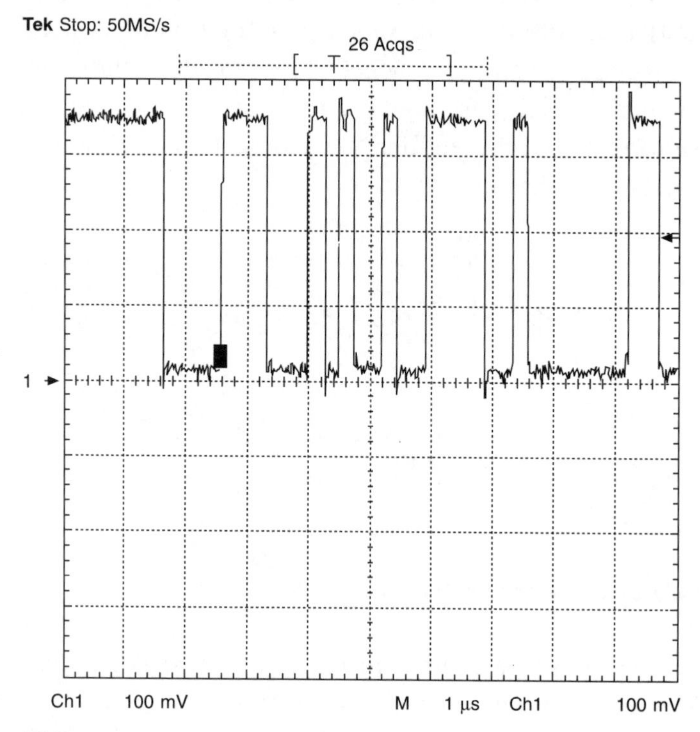

Figure 10.3

Testing processing chips

The absence of an output from a processing chip does not necessarily mean that the chip is faulty. In a microprocessor-based system, IC output failure may be a result of one of several malfunctions – including those affecting a clock pulse, control signal or software routine, data or address bus lines, a memory chip, or a DC supply voltage and supply lines – as well as of a faulty IC itself. In general, before a suspected IC is replaced it is important to check that it is:

- Receiving its DC voltage supply voltage at the appropriate pins; if not, then the DC line should be traced back to the power supply to ascertain the fault
- Operating at the correct frequency, which may be checked by a suitable oscilloscope. In some cases, a drift of as little as 10 per cent may cause a microprocessing chip to malfunction. Where more than one clock pulse is fed into the chip, they must all be checked in the same way
- Receiving the correct instructions from the microprocessor or the microcomputer, or both. This involves checking the control signals, such as the CS, IRQ and R/W signals, all of which should display digital activity of some sort. It may also include checking the series control bus lines and data and address lines for digital activities.

The DC voltage check may be carried out using a DVM to measure the DC voltage at the appropriate pins of the chip. The clock signal may be checked using an oscilloscope with adequate frequency range. Apart from the frequency, the clock pulse must have a fast-rising square shape with, normally, an amplitude of around 4.5 V.

Assessing hardware control lines involves using a logic probe to check digital activity, or a storage oscilloscope to examine the actual waveforms of the various control signals such as chip select (CS), acknowledge (Ack), READ, WRITE, strobe and interrupt. The *Reset* (*RST*) pin is one of the few lines that has no digital activity present; the others, being V_{CC} and 0 V, may be tested by a logic probe or a DVM. RST is normally active LOW, i.e. it resets the chip when it is taken LOW and restarts when taken HIGH. If it is active LOW, then the RST line should be permanently HIGH. The signals at the data and address lines may be examined, using a logic probe or oscilloscope, for digital activity.

Faulty software in terms of out-of-date, deleted or corrupted programs may also cause an interruption of the bitstream. Software processing routines are mainly stored in Flash memory chips.

In microprocessor-based systems, failure may also occur due to wrong or missing clocks, or incorrect timing of address, data and control signals. A fault in timing or synchronization can result in partial or total failure of the system. The timing and synchronization of the various data and control bitstreams may be examined using a multi-trace oscilloscope, which displays two or more signals simultaneously. A typical multi-trace display illustrating the time relationship between chip select (CS), READ or output enable (OE), a data line, and the clock of a memory chip is shown in Figure 10.4.

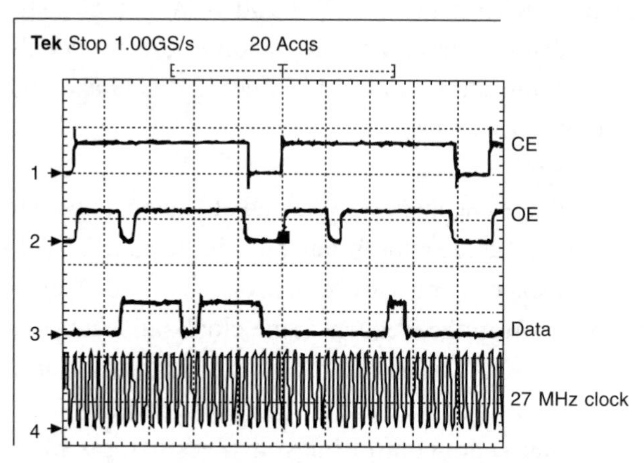

Figure 10.4
Multichannel digital storage scope: typical display

Analogue signals

While most test signals in a DVD player are digital, analogue signals are also present – mainly at the front end and the output ports. Testing these signals for amplitude, shape and frequency is crucial to any servicing procedure. Apart from the RF signals from the optical head, which produce the familiar eye pattern on an analogue oscilloscope (Figure 10.5), the other analogue signals

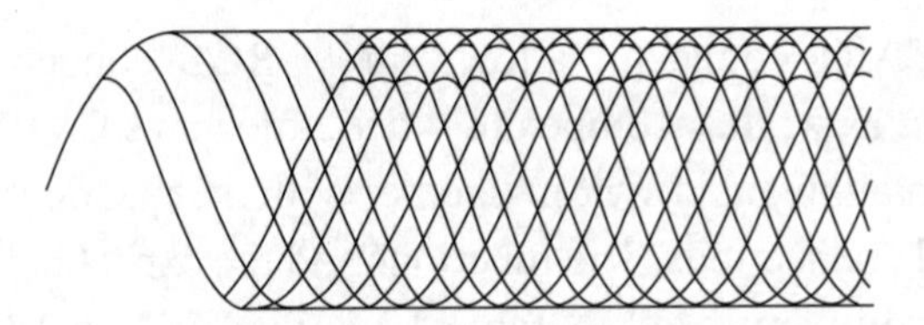

Figure 10.5

are the video and audio waveforms at the output port. One of the most convenient points to observe the presence of these V/A signals is at the appropriate pins of the SCART and the S-video connectors.

The SCART connector

One of the standard outlets for video and audio information is the SCART socket connection (Figure 10.6). The SCART connector is a 21-pin non-reversible device. There are 20 pins available for connections, with pin 21 connected to the skirt and hence the chassis to provide the overall screening for cable communications.

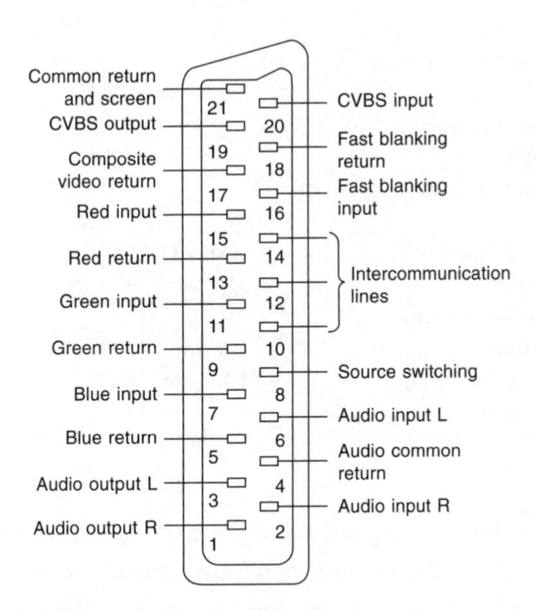

Figure 10.6
SCART socket pin-out

The SCART connector allows direct RGB connection to a TV receiver, as well as composite video, known as CVBS (composite video, blanking and sync), together with independent stereo sound channel connections. The functions and the expected level of signal on each pin are listed in Table 10.1. Pins 10 and 12 are used for intercommunications between devices connected to the SCART socket.

Table 10.1 SCART socket pins and their functions

Pin	*Function specification*	*Signal*
1	Right channel audio out	0.5 V into 1 kΩ
2	Right channel audio in	0.5 V into 10 kΩ
3	Left channel audio out	0.5 V into 1 kΩ
4	Audio earth	
5	Blue earth	
6	Left channel audio in	0.5 V into 10 kΩ
7	Blue in	0.7 V into 75 Ω
8	Source switching (9–12 V)	Not specific, but usually max. 12 V into 10 kΩ
9	Green earth	
10	Intercommunication line	
11	Green in	0.7 V into 75 Ω
12	Intercommunication line	
13	Red earth	
14	Intercommunication line earth	
15	Red in	0.7 V into 75 Ω
16	Fast RGB blanking	varies (1–3 V)
17	CVBS earth	
18	Fast blanking earth	
19	CVBS out	1 V into 75 Ω
20	CVBS in	1 V into 75 Ω
21	Socket earth	

For a standard test signal, namely the colour bar display shown in Figure 10.7a, the expected waveforms at each of the video outputs (namely CVBS, Red, Green and Blue) are illustrated in Figure 10.7b.

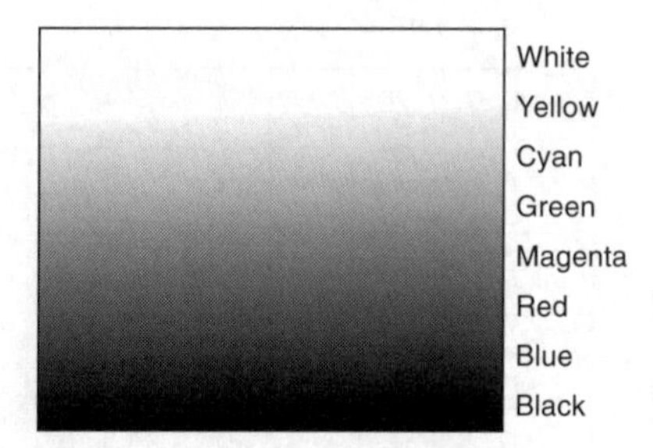

Figure 10.7a

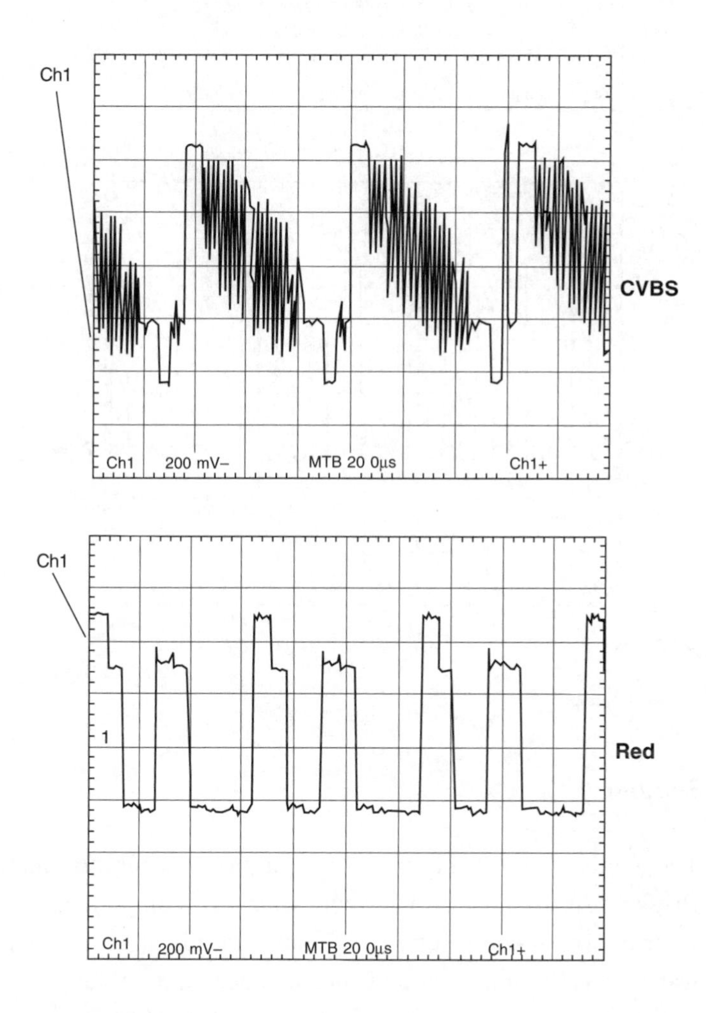

Figure 10.7b ***(Contd)***

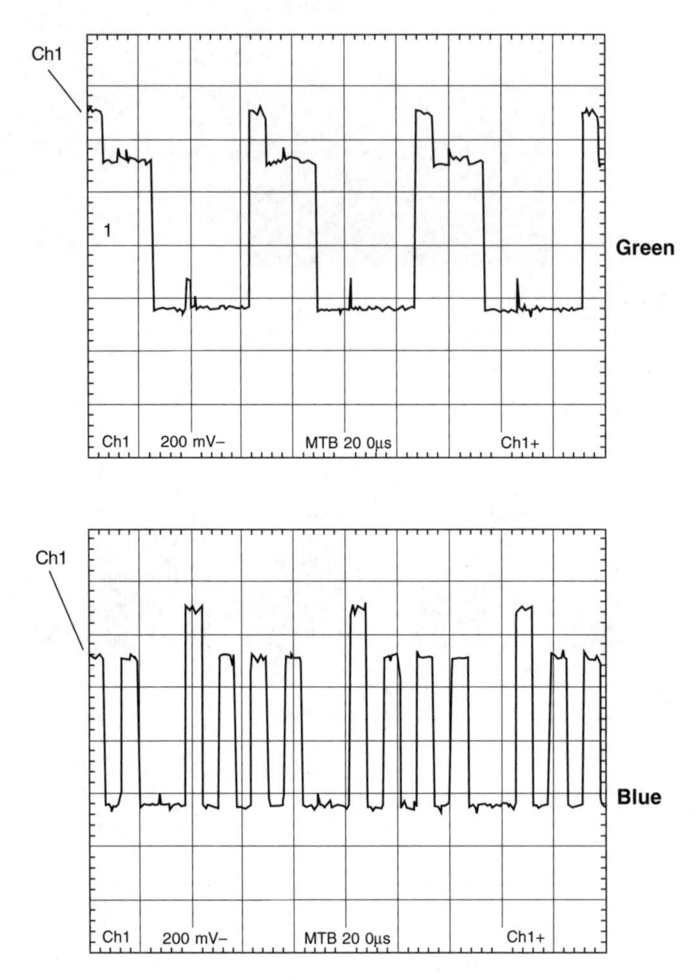

Figure 10.7b

The S-video

The S-video socket (Figure 10.8) is a more recent introduction to British V/A appliances. It provides the two components of video information, namely luminance and chrominance. The expected waveforms (luminance Y and chrominance C) for a standard colour bar test signal are illustrated in Figure 10.9.

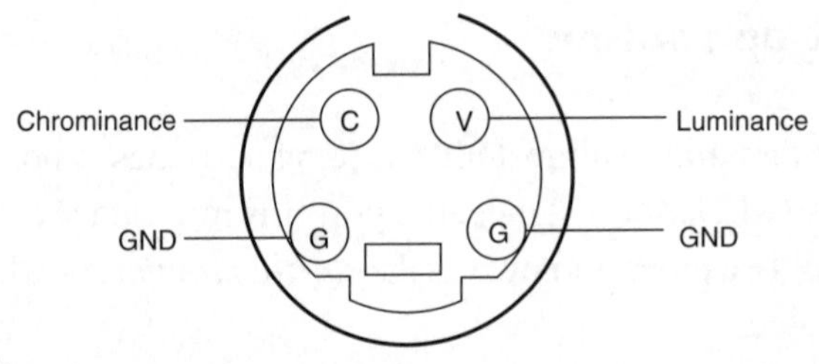

Figure 10.8
S-video socket

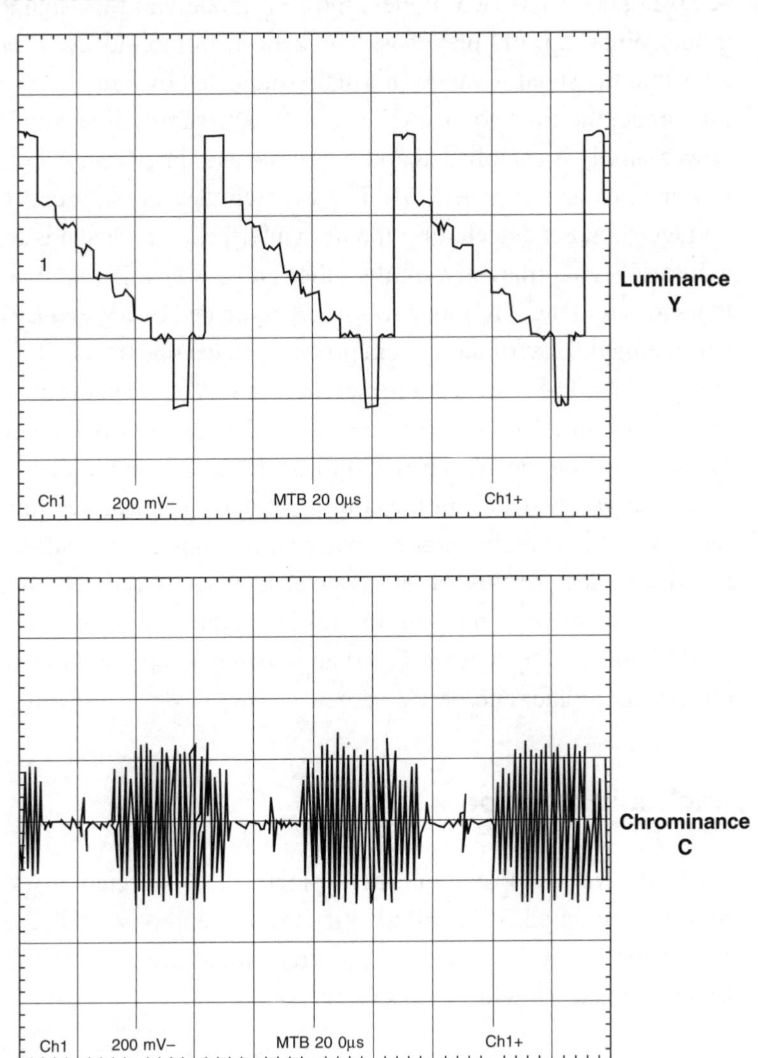

Figure 10.9

The start-up routine

A starting point in almost all servicing activities is observing what happens when, for instance, the power is recycled (i.e. switched on from cold), a process known as the *start-up routine*, or when playback is initiated.

A DVD player has two modes: the 'on' mode and the 'standby' mode. When a DVD player is switched on from cold, it initially goes into the standby mode (normally indicated by a red LED). In this mode, the ever voltage (E, e.g. 3 V) DC supply line from the power supply is established and fed to the user interface. All other power lines are switched off. The ever voltage or a special OK voltage signal is detected by the user interface, which sends back a control signal to switch on the other DC power rails; the player then goes into the 'on' mode (normally indicated by a green LED), energizing the rest of the system processors, decoders and all other components. The system controller then carries out a set of routines to test and initialize all the other units. This process is known as the start-up routine. When the start-up routine is completed, the user interface switches off the control signal and the power supply returns to the 'standby' mode – where it remains until a button on the front panel or the remote control handset is pressed. If the standby/on button is pressed, the DVD player will go into the on mode (indicated by a green LED) and display a logo or a message on the front panel and/or TV screen.

Playback initialization

When the tray is closed, or when 'playback' is selected, a *set-up* routine is initiated. This will identify the size and type of disc, and set up the two-axis actuator for correct focus and tracking. The following actions will be observed:

- The spindle motor rotates the disc at a relatively high speed.

- The laser beam is directed onto the disc, and the reflected beam is detected by the photodiodes.
- The sled motor moves the optical head away from the centre in one or two steps while the two-axis actuator simultaneously moves the objective lens up and down to obtain a focus – a process known as *Focus Search*. The movement across the disc produces a tracking error. If the tracking error voltage is low (0.4 V or less), then the inserted disc is a DVD. A CD disc, with its longer pits and wider track pitch, will produce a higher TE voltage of around 2 V.
- During the focus search, the strength of the reflected laser beam is monitored to produce a pull-in (PI) signal. A high PI signal of around 1 V indicates a highly reflective surface, indicating a single-layer disc. Alternatively, a low PI of around 0.5 V indicates a dual-layer DVD disc.
- If focus is achieved, the optical head moves to the inner circle to read the first sector and produce an RF signal. The bit rate of the RF signal is used by the RF processor to set the spindle motor to the correct speed, and playback commences.

Common symptoms and their possible causes

1. *Spindle motor speed goes up and down several times and then stops; 'No disc' is displayed.*
   - These are classic symptoms of having no RF signal, caused by several malfunctions – including focus failure, tracking failure or optical head failure. Use self-diagnostic routines to identify the location of the fault in terms of focus, tracking or laser power. In any case, the FE, TE and RF signal paths may be examined from the RF amplifier right through to the two-axis actuator drives during the period when the spindle motor is turning and the optical head is searching for a focus. The laser beam and the two-axis actuator should also be checked. Before suspecting faulty chips, assess the continuity of the relevant fusible resistors.

2. *Spindle motor ticks few times in an attempt to turn and then stops.*
   - These symptoms indicate a spindle or sled motor drive failure. Check error signals and motor drive signals.
3. *Stuck at standby.*
   - This symptom can be a result of a partial failure of the power supply (no switched DC lines), or of having only ever voltage lines. The start-up sequence will not take place, and the power LED will remain at RED when AC power is recycled. The first step is to determine whether the fault lies in the power supply unit (PSU) or outside it. In some power supplies, the power supply may be forced into the 'on' mode by over-riding the control signal from the user interface by connecting the ever 3-V supply line (via a 100–500 Ω resistor) to the 'control' signal. If the PSU remains in the standby mode, the fault lies within the PSU itself. If, on the other hand, the power supply switches into the on mode, the fault is outside the PSU.
   - Stuck at standby could be a result of failure of one of the processing or decoding chips (RF processor, A/V decoder). The sys con processor chip will detect such a fault during the start-up routine (or when a function is selected by the user), switching the power supply into standby with the control signal via the user interface. Stuck at standby could also be caused by a malfunctioning user interface or sys con processing chips.

Power supply failure symptoms

A number of power lines are provided by the switched-mode power supply. Typical DC power lines are:

Ever 3 V	For user interface
Switched 3 V, 5 V	For optical unit and processing chips, such as RF amplifier, RF processor, A/V decoder, system controller, memory chips and servo control
Switched +DC, –DC	Higher voltages (in the region of +10 V and –10 V) are necessary to drive the spindle, sled and tray open/close motors, front panel control switches, front panel display, audio output section and SCART switching

Symptoms of a PS failure vary from one DVD player to another, depending on the power supply circuit design. Typical symptoms include the following:

1. *Non-rotating disc, no video, no audio.*
   - This may be a symptom of total/partial PS failure. See if the player reverts to standby; if so, then check for partial power supply failure. Otherwise, check for total failure. Check the output DC supply lines, and the fuse on the primary side of the transformer. Check the various voltages (ensure that an isolation transformer is used) to detect the actual faulty component.
2. *Fast-rotating disc, no video, and no audio.*
   - This may be a symptom of failure in one or both high DC voltages (+10 V, –10 V). Check whether the player reverts to standby. If this is the case, it indicates possible partial PS failure. Check the high DC voltages. If present, then check for a break in the path.
3. *Normally-rotating disc, no video, no audio.*
   - This may be due to partial PS failure. Check whether the player reverts to standby. If this is the case, it indicates possible partial PS failure. However, this will depend on the circuit design. Over-ride the ‘control’ signal to switch PSU into the on mode, and check DC output lines. If all DC output lines are correct, the A/V decoder should be suspected.

DVD self diagnostics

Modern DVD players provide a comprehensive software routine for self-test and adjustment. The routine is executed by the system control microprocessor, with test and adjustment options as well as results of tests displayed on the screen. It allows the user to set up the player, and check the servo functions (including focus and track), the spindle and tray motors, and the memory chips. Failures will be indicated on the screen, sometimes with special codes to specify particular faulty components. It should be noted that the self-test routine can only be used if the power supply, system control, user interface, remote control and front panel, and video displays are functioning normally.

Chapter 11

Data flow

Introduction

As described in Chapter 5, data are constructed and written onto the disc in blocks of 4836 bytes. The optical head then reads these blocks off the disc, one by one, and the resulting channel data bitstream is fed into the DVD player for processing and control. The physical organization of the data on the disc is laid down by the DVD standards, which specify that data must be placed sequentially (physically contiguous) in accordance with the video structure, starting with the *video manager* (VMG), which contains the main menu. Normally a DVD disc has a single film and hence a single main title. The main menu is sometimes known as the *title menu* or *top menu*. The VMG is followed by the *video title set* (VTS), which contains information about each title listed in the main menu. The structure is hierarchical in that the VMG and VTS blocks are broken up into sub-blocks, which are in turn broken up into further sub-blocks, and so on.

Data flow

The optical head placed under the disc surface will read the data, block by block, as the disc rotates. Data are read at a constant rate of 26.16 Mbits/s. In order for the data to be read at a constant rate, the disc must rotate at what is known as constant linear velocity (CLV). This means that when the optical head is reading a track nearest to the inner circumference the disc rotates at an angular velocity of 25.5 revolutions per second, compared with 10.5 revolutions per second when it is reading a track nearest to the outer circumference.

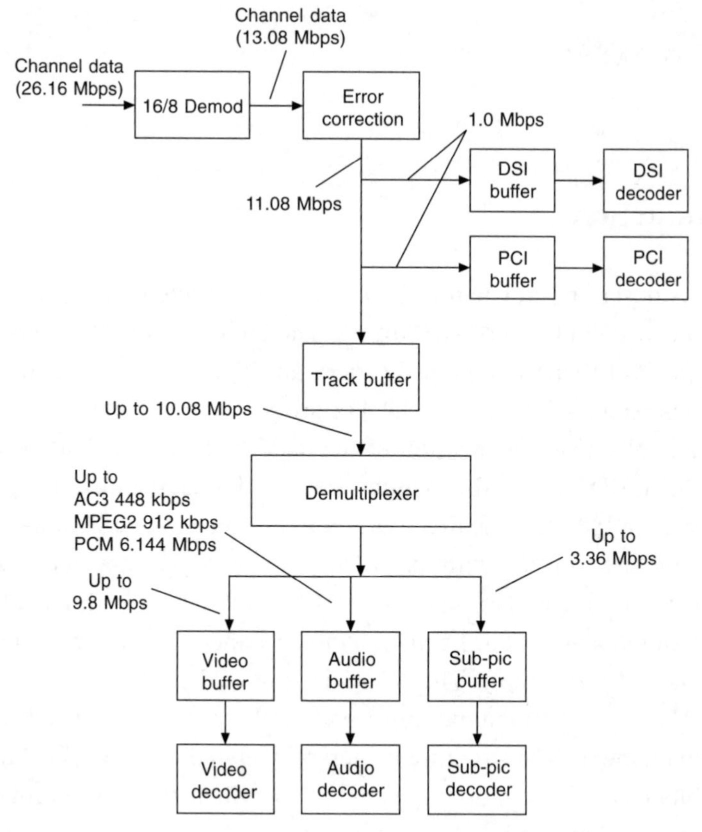

Figure 11.1
Data flow (version 1)

Figure 11.1 shows the flow of data in DVD players. Channel data from the optical head contain five different packetized elementary streams (PES):

1. Video PES (one stream)
2. Audio PES (up to eight streams)
3. Sub-picture PES (up to 32 streams)
4. PCI (presentation control information)
5. DSI (data search information).

Having entered the processing section at a constant bit rate of 26.16 Mbits/s, the constant data flow rate is reduced by 8/16 demodulation to

$$26.16/2 = 13.08 \text{ Mbits/s}$$

Following error correction, the error bits are removed and the data flow rate is further reduced to 11.08 Mbits/s. At this stage DSI is copied into a buffer at 1 Mbits/s for a special encoding process. The remaining components of the data block are then fed into a track buffer at 11.08 Mbits/s before going into the individual video, audio, sub-picture and PCI buffers via a demultiplexer. While the video, audio, etc. data enter the track buffer at a constant bit rate, they are fed into their respective buffers at a variable bit rate. Variable transfer rate allows for the efficient use of the total transfer rate by allocating it to where it is needed at any one time. For the same image quality, a constant video transfer rate would be twice that required using a variable transfer rate. Video CD uses a constant video transfer rate of 1.15 Mbits/s; hence the low picture quality.

Within the 10.08 Mbits/s constraint on the track bit rate, each individual PES is limited to a maximum bit rate as shown in Table 11.1. While video streams normally use a variable bit rate, DVD audio streams generally use fixed transfer rates. The audio bit rate is usually set to the maximum permitted bit rate, as listed in Table 11.1.

Multiple-angle feature

In a DVD disc with the multiple-angle feature, pictures from up to nine different camera angles may be pre-recorded on the disc and

Table 11.1 Maximum permitted bit rates

Video	9.8 Mbits/s
Audio PCM	6.144 Mbits/s (eight channels)
Audio Dolby Digital	448 kbits/s
Audio MPEG-1	384 kbits/s
Audio MPEG-2	912 kbits/s
Sub-picture	3.36 Mbits/s

the viewer may switch to the desired camera angle in real time during the playback process. For example, in software that stores a live football match, the pictures from behind one or both goals can be pre-recorded; in the case of a concert, pictures of the vocalist, guitarist, etc. can be pre-recorded on an individual basis in addition to the normal recording. The viewer may then select any of the available pictures, offering an entirely new experience in visual entertainment. The essential point of the multiple-angle feature is its capability to play back simultaneously-occurring multiple series of image frames without showing any joints. This type of playback is known as *seamless*.

Seamless playback

After reading video, audio and other relevant packets from the disc, the DVD player stores this information into a track buffer memory store. Each packet contains a *packet identifier* (*PID*) as well as a *time stamp*; the PID specifies the program stream, while the time stamp provides the necessary synchronization between video and audio packets (lip sync) as well as determining the correct sequence of video packets. Where the multiple-angle feature is available, the data recorded on the disc contain packets belonging to up to nine different angles. The pickup head reads only the necessary portion of the data on the disc that is associated with the selected program, while jumping over irrelevant tracks. The process of playback involves retrieving the various packets with the relevant PID from memory in the order specified by their time stamp – i.e.

in the order in which they are to be displayed. This is called data buffering, and employs a 16-Mbit memory chip.

If during playback a different angle is selected, the pickup head will be directed to read the different portion of the disc that contains the data packets for the selected angle, and store them in memory. While the new data are read from the disc, packets from the memory buffer continue to be displayed so as to accomplish continuous image reproduction without leaving any visible joints or seams. The size of the buffer depends on the actual data structure on the disc, although a minimum of 2 Mbits of memory is recommended; this accommodates a maximum jump distance of 10 000 sectors. In some DVD players the track buffer is combined with other memory chips, such as the V/A decoder memory or the video buffer.

Multi-story and parental lock

Other playback facilities using the seamless capability of DVD are the multi-story function and the parental lock. In the *multi-story* function, the story line may be determined by the user from a selection of options on the screen. The operation of this function is essentially the same as that for multiple-angle function.

Parental lock or parental control allows the 'locking' of software at the discretion of parents to prevent explicit violence, horror, sexual activity or any other undesirable clips being viewed by children. For this function to be available for the viewer, the DVD disc must contain the necessary software to skip any undesirable scenes automatically and switch over to alternate scenes when the parental lock function is activated (Figure 11.2).

Disc capacity and transfer rate

Although DVD storage capacity is significantly greater than that of audio CDs, it is still limited to 4.7, 8.54, 9.4, 13.24 or 17 GB,

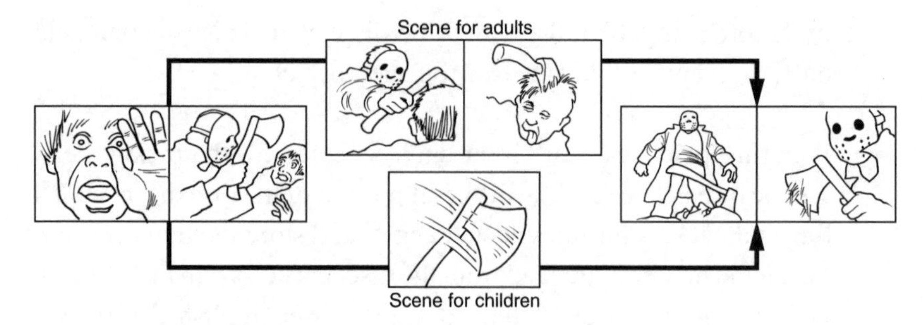

Figure 11.2
The parental lock function

depending on the number of recorded layers and sides. The capacity of the disc determines how much video and audio information can be stored on a given disc. In 8/16 modulation each 8-bit word is converted to a 16-bit word, thus doubling the number of bits representing the same amount of information. This would effectively halve the capacity of the disc if it were not for the technique used in writing the data on the disc. This technique, the non-return-to-zero inverted (NRZI) technique, halves the amount of transitions required to record any given number of bits, and effectively doubles the capacity of the disc.

The stored data are transferred from the disc to the player at an average rate, which will determine the total playing time of the disc. For instance, a 4.7-GB (billions of bytes) disc with an average transfer rate of 3.5 Mbits/s will result in total playing time that may be calculated as follows:

$$\text{Capacity} = 4.7\ \text{GB} = 4.7 \times 10^9\ \text{bits} \times 8\ \text{bits/byte}$$
$$= 37.6 \times 10^9\ \text{bits}$$

(As discussed above, 8/16 modulation effectively halves the capacity available for recording. On the other hand, the NRZI doubles the capacity, leaving the recorded capacity the same – namely 4.7 GB or 37.6×10^9 bits.)

$$\text{Transfer rate} = 3.5\ \text{Mbits/s} = 3.5 \times 10^6\ \text{bits/s}$$
$$= 3.5 \times 10^6 \times 60\ \text{bits/min}$$
$$= 210 \times 10^6\ \text{bits/min}$$

$$\text{Playing time} = \text{capacity/average transfer rate}$$
$$= 37.6 \times 10^9\ \text{bits}/210 \times 10^6\ \text{in minutes}$$
$$= 179\ \text{minutes}$$

Conversely, if the playing time is set at, say, 150 minutes, and recorded capacity is known – say 4.7 GB – then the average transfer rate may be calculated:

$$\text{Average transfer rate} = \text{capacity/time}$$
$$= 4.7 \times 10^9 \times 8\ \text{bits}/150 \times 60\ \text{s}$$
$$= 4.7 \times 10^3 \times 8\ \text{Mbits}/150 \times 60\ \text{s}$$
$$= 37.6 \times 10^9\ \text{bits}/9000\ \text{bits/s}$$
$$= 4.178 \times 10^6\ \text{bits/s}$$
$$= 4.178\ \text{M bits/s}$$

Presentation, navigation and search

A DVD disc contains two types of information; control information for *navigation*, and user data for *presentation*. A complex set of navigational tools is available to provide the necessary flexibility to control the presentation of video pieces and provide such facilities as 'search', parental control, and multiple angles. Playback is controlled by a number of navigation tools that provide menus, seamless branching, parental control, multiple angles, and low-level interactivity. Search facilities are provided by a variety of buttons, including 'next', 'previous', and 'fast forward'.

The basic unit of presentation is the *cell*, which is the smallest addressable chunk of information. A film or part of a film (a chapter or section) is a sequence of cells. Each cell has a unique ID number, which can be called by a *program* (*PG*) within a *program chain* (*PGC*) for decoding and presentation. DVD supports up to 999 PGCs for every title, and 99 PGs within every PGC.

Figure 11.3 shows the structure of a single monolithic title or film. It has a single PGC and the playback is linear (hence the term monolithic), with no facility for branching to change the order of the cells.

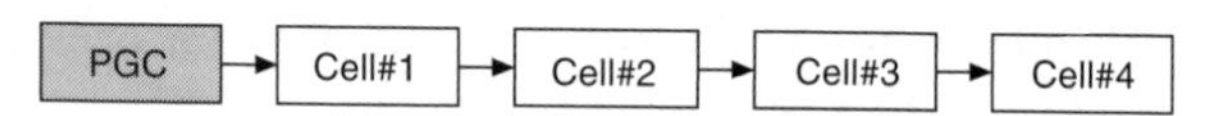

Figure 11.3
Single-title monolithic playback

Most DVD video applications provide multiple program chains, and the sequence of cells will thus depend on the PGC selected. Figure 11.4 illustrates the structure of a DVD containing two titles, with one PGC for each title.

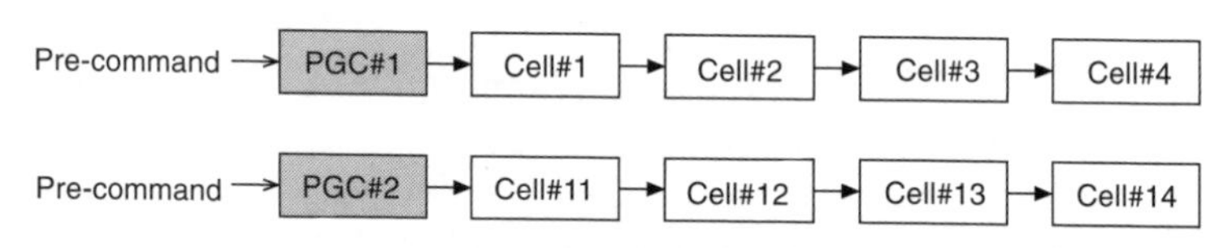

Figure 11.4
Multiple-title, monolithic playback for each title

Normally, branching is necessary to provide such facilities as parental control (where some scenes, in the form of cells, are blocked or bypassed), multiple angles (which allow the viewer to select a different camera angle for the same scene) or multiple stories (which allows the user to select a storyline or a story ending). In these cases, the player is directed to move from one program chain to another by a special command without any interruption to the displayed video – i.e. seamless playback. Individual cells may be used by more than one program chain (Figure 11.5). Generally a

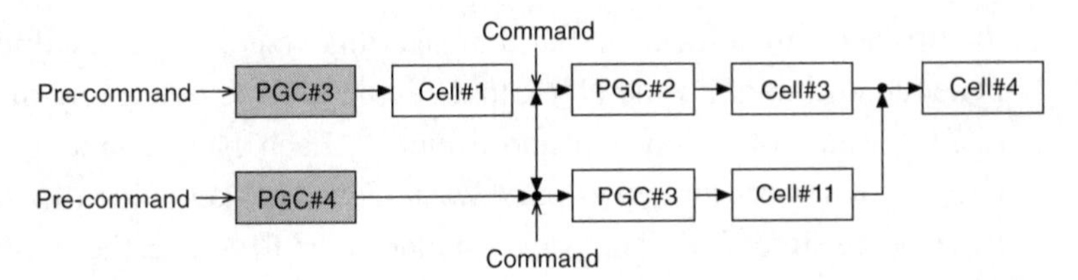

Figure 11.5
Branching

program chain consists of a series of program chains and so on in a complicated navigation system.

The cell-structured data

A cell may be as short as 1 second, or as long as a complete film. It consists of a whole (integer) number of group-of-pictures (GOP) or audio blocks in the form of one or more *Video Object Units* (VOBU; Figure 11.6). Each VOBU contains a GOP with short-duration (0.4–1 s) video, audio and other data in the form of *packs*. One of the packs, NV_PCK, is dedicated to navigation information, which defines the behaviour of the cell. Each pack has a header that, among other things, contains a 6-byte (48-bits) *system clock reference* (SCR) code. This code represents a 90-kHz sample of

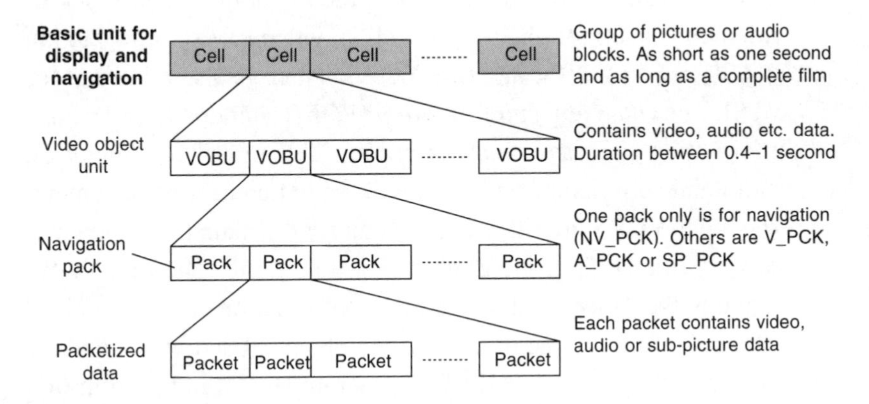

Figure 11.6
Cell structure (version 2)

the original time clock used at the encoding stage, and is used to generate an internal time clock that can be used to determine the display times of video and audio pieces. Each pack contains a number of sectors known as *packets*, and each packet is identified by a 1-byte stream ID. The packet includes the PES, together with error correction and sync data.

A video stream is a series of group of pictures (GOPs) in the form of picture frames, not necessarily in the correct display order. The frames themselves are a series of relatively small size packetized elementary streams (PESs), simply known as packets. Playback involves the selection of groups of PESs in a particular order, and this is carried out by '*pointers*' under the control of the video manager.

Video manager

Data are written on the disc starting with the video manager (VMG), which contains the main menu. The main menu lists all available titles on the disc. A title is a complete feature, such as a movie. Normally there is just one title, in which case the disc may go into autoplay, displaying the options available under that title. Each title is contained in the next block of data recorded on the disc, namely the *video title set* (VTS). The video manager may contain up to 99 video title sets, each of which contains three types of video object sets (VOBS; Figure 11.7); video title set information (VTSI), menu VOBS, and title VOBS. There is also a backup for VTSI. The *video object information* (*VOBI*) contains video control information such as the aspect ratio, PAL or NTSC, language, audio and sub-picture options, and parental control management. The menu VOBS provides a sub-menu for the main title, known as Part_of_title (PTTP) – e.g. chapters, trailer, etc. The title VOBS contains the video, audio and other information for the selected title in the form of one or more video objects. Each VOB contains part of or the complete MPEG-2 program stream in the form of a number of cells. Video objects within a video object set are identified by an individual ID number.

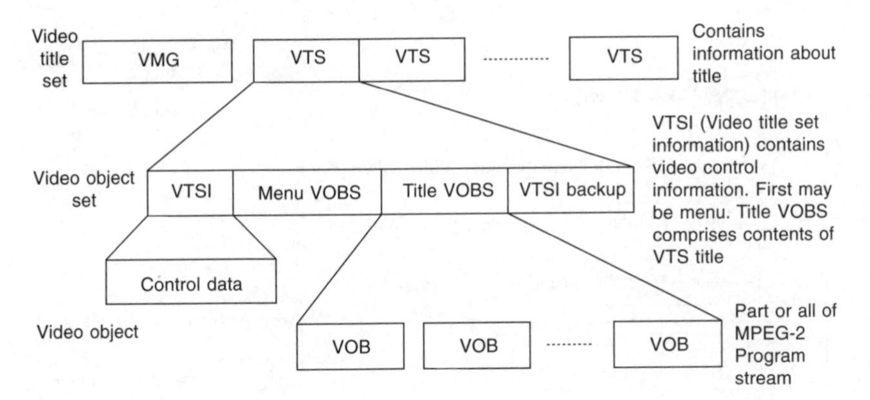

Figure 11.7
Video manager

Volumes and directories

The DVD disc is basically a high-capacity storage device not dissimilar to a computer hard disk. Just like a hard disk, data on it must be organized in a format that a computer system can access – i.e. a file format – and the file format chosen is the UDF file format. However, unlike a hard disk or a DVD-ROM (which require a versatile and comprehensive file format) a DVD-video is a dedicated disc with fewer needs in terms of file structure. For this reason, and in order to limit the computing power of a domestic DVD player, a number of constraints are added to the UDF format, including limiting the disc to one volume, one partition and one file set. DVD-video file specification describes a set of files for video and audio information. All the video files are stored in a directory called Video_TS, and all the audio files are placed within a directory called Audio_TS. The directory structure of a single title disc is illustrated in Figure 11.8.

DVD recording

The commercial success of DVD players and drives has spawned an obvious successor, namely the DVD writer. There are three distinct types of rewritable formats: DVD-RAM, DVD+RW and

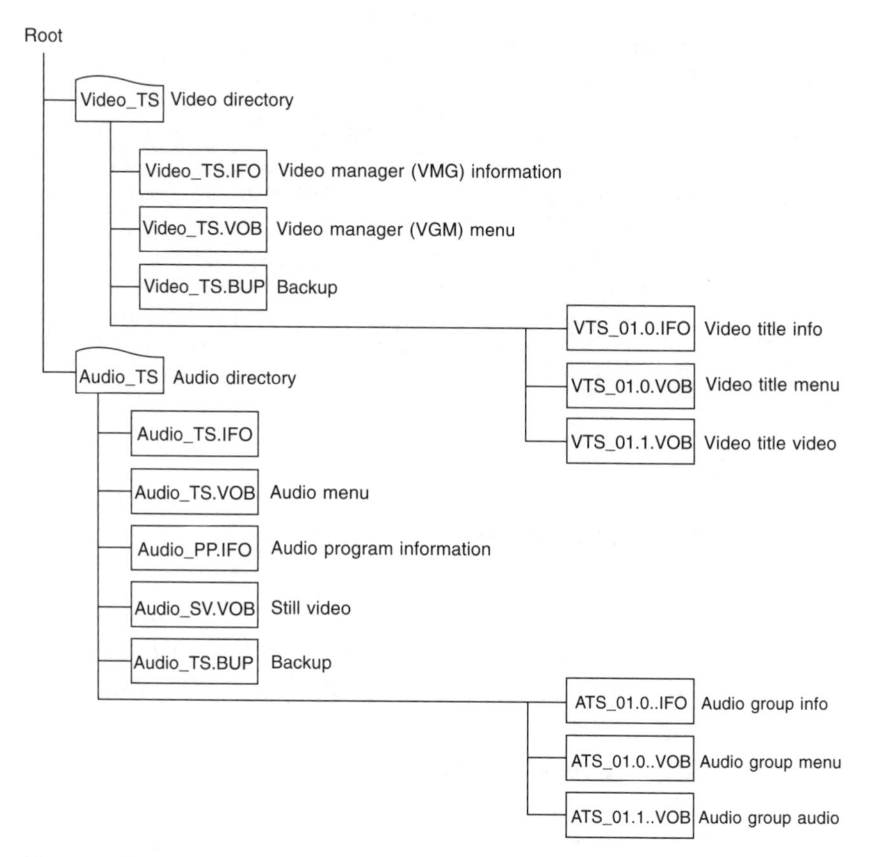

Figure 11.8
Directory and file organization for a single title

DVD–RW. All three DVD rewritable formats have a stated capacity of 4.7 GB per side, and all three formats provide the user with a facility for selecting the data transfer rate (or bit rate) of the recording. While higher bit rates yield better quality recordings and hence better quality video and audio, they eat up more of the disc space resulting in a shorter recording time. Both DVD+RW and DVD–RW allow the user to adjust the bit rate on the fly. The user can preset the desired bit rate level, but the recording machine will examine the source material (e.g. the video from a TV film) and will raise the bit rate to ensure that, for instance, an explosion does not result in a splotchy display when played back. On the other hand, the recorder will scale back the bit rate in other, less demanding scenes, so that the bit rate remains at the average value set by the user. DVD-RAM will do the same thing if set to the variable bit rate mode. However, DVD-RAM can also be set to the constant bit rate mode, in which case the bit rate will be the same regardless of the content of the video.

Some DVD writers provide a facility, supported by DVD+WR, for constant angular velocity (CAV) recording as well as constant linear velocity (CLV) recording.

CHAPTER 12

DVD PRODUCTION

Introduction

The production of a traditional film on a video tape, or audio tracks on a CD, is relatively straightforward in that the primary purpose is the creation of a pre-master, which may then yield several copies that are as close in quality to the original source as possible. Very few artistic or production skills are necessary. However, this is not the case with the production of DVD titles. With the advent of DVD-video the preparation of commercial titles has become far more complex, requiring new skills in allocating resources and in menu and navigational design. In audio applications, DVD offers many new options in the preparation of music titles. Music tracks may be presented not only in stereo, but also in one of many variations of multi-channel surround sound. They may be presented as audio-only pieces, or with text, graphics or motion video. DVD production involves multiple media and multiple presentation settings to create a new experience for the viewer.

DVD production involves:

- Planning
- Assets preparation
- Resource allocation

- Authoring
- Formatting.

Planning involves determining the basic structure (menus, navigation, etc.), elements and scope (audio, video, etc.) of the DVD title. Following planning, the sources from which the various elements are to be provided have to be ascertained and prepared. Once a list of the assets has been drawn, the resources required to implement the desired structure and scope of the DVD title have to be calculated to ensure that they fall within the specified limitations of the DVD disc. If the resource requirements are greater than the available resources, the structure and/or scope of the title and consequently the assets will have to be modified. This is then followed by the actual allocation of resources to each aspect of the title (e.g. audio, video, subtitles, sub-pictures, etc.).

Authoring is the next step in the production process, and involves the integration of the various media elements into a unified whole with navigational pathways linking the various parts. Interactive elements of the project are also included at this stage. The final step is the physical formatting of the disc and subsequent testing to ensure a high-quality product.

Planning

Planning is the first stage of the production of a DVD title, and involves making decisions on:

- The type of DVD playback platform the disc is intended for. A DVD disc may be designed to play in any one of several playback platforms – such as a DVD-video/audio player, DVD-audio only player, or PC-based DVD drive – or it may be intended for all of the available platforms. The choice of the platform will set the available options, as well as the limitations that will constraint the production.

- The general feel of the DVD in terms of menus, buttons and background to menus. For instance, menus with motion video background and elaborate sound provide a more exciting environment than menus with background with still pictures, but they require more disc space and time-consuming authoring processes. On the other hand, a less flashy menu environment may do the same tasks in introducing the user to the disc contents whilst saving on production time.
- The level of interactivity to be included in the title. This is particularly important in DVD-video games, although it may arise in other productions to a limited extent. The more interactivity is included, the more elements it uses and the more disc space is required.
- The amount of video to be included in the title.
- The type of audio to be made available. The range of audio options available in DVD technology is very diverse, and includes stereo and a variety of surround sound applications. A DVD title may include any combination of these options.
- The extent and type of text/subtitles to be made available, including the multilingual text option. Each additional language used may involve translation and synchronization with the audio and video content.
- Including an additional language audio track. Additional audio tracks involve translation, editing and synchronization, all of which increase the cost of the production and utilize disc space.

As an example, consider a DVD-audio title designed for both the UK and French markets. The title is made up of two parts: a 75-minute live video concert recording with Dolby Digital sound, and a 60-minute music section containing 12 tracks of high-quality PCM surround sound. The music section is accompanied by still images and text. The elements of the title are therefore:

Audio Dolby Digital	Six tracks (5.1) of 75-minute video soundtrack
Audio PCM	Six tracks of 60 minutes audio 24-bit/96 kHz
Video	60 minutes of MPEG-2 video
Stills	Specified number of stills
Menus	Specified number of background stills for menus
Text (subtitle)	60 minutes of English lyrics
Text (subtitle)	60 minutes of French lyrics

Assets preparation

Once the planning decisions have been taken, a title template or a storyboard can be constructed setting up the organization of the title. From this, a list of the *assets* may be derived. A typical title project requires a few dozen assets from a variety of sources, including video, audio, graphics and subtitles. Each asset must be adequately prepared before it can be fed into the DVD production process. The final quality of a DVD title is primarily dependent on the quality and visual character of the source material, and the bit rate allocated to the various elements of the title.

Video assets

Preparing the video assets involves ensuring that the video is of a good quality digital source. An analogue source must be converted to the digital format before it can be used. If the video is of poor quality or contains noise, then *digital video noise reduction* (*DVNR*) must be used to improve it. Noise may be a result of grainy film, dust or video snow, or it may be satellite noise – in all cases it is random unwanted signals. DVNR compares the video across a series of frames, and removes any random noise present.

At this assets preparation stage, some parameters of the MPEG-2 encoding process have to be set. As we know, video encoding involves two types of data compression: temporal and spatial. Temporal data compression starts with a reference frame, known as the I-frame, with which subsequent frames are compared. A motion vector is produced as well as a predicted difference frame, known as a P-frame. This continues for a number of frames known as a group of pictures (GOP), after which the process starts again with a new reference I-frame and so on. The precise number of frames in a GOP has to be determined by the DVD producer, and is typically 10 or 12 frames. However, this may be changed as necessary to suit the particular video sequence. The number of frames in a GOP need not be a fixed figure throughout the video program. For instance, whenever a cut is introduced and the scene changes from, say, an indoor set to an outdoor location, a new GOP has to be started with a new I-frame. The cut may not coincide with the end of a GOP, making it necessary to end the GOP prematurely. A smaller GOP is also used where there are large differences between successive frames – for instance, in cases of fast-moving content.

A more efficient technique in terms of data reduction is the use of bidirectional prediction frames, B-frames. These frames are produced as a result of comparing a frame with the following as well as the succeeding frame. For a given picture quality, a coded I-frame needs three times more bits to describe it than does a coded P-frame – which itself requires 50 per cent more bits than a coded B-frame. For this reason, a typical group of pictures will consist of an I-frame followed by a number of P- and B-frames (Figure 12.1). The precise combination is determined by the DVD producer.

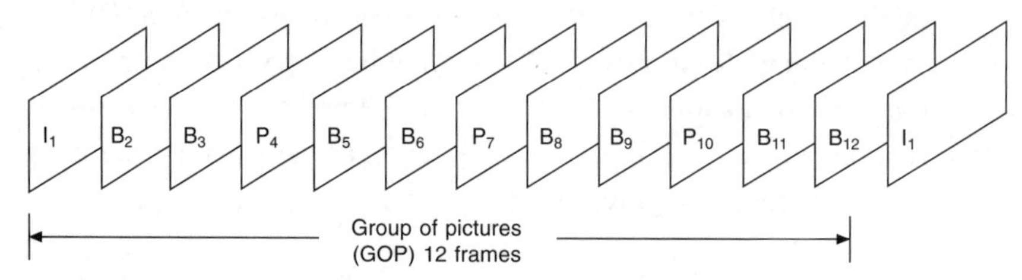

Figure 12.1
GOP with M = 12 and N = 3

The composition of the group of pictures is described by two parameters; the number of pictures in the group M, and the spacing between anchor frames (I or P), N. For instance, in the group of pictures illustrated in Figure 12.1, M = 12 (12 frames in total) and N = 3, indicating that an anchor frame, I or P, occurs every third frame.

Audio assets

Preparing the audio assets may be as complex or even more complex than preparing the video assets. The quality of the source material is obviously a crucial factor in the ultimate quality of the audio element of the title. However, unlike video, audio has a number of tracks – two for stereo, six for Dolby Digital 5.1, and eight channels for PCM audio. All of these tracks must be synchronized with one another and with the other elements, including the video element. Furthermore, each track must have its level adjusted to establish consistency of the multi-channel sound. It is important to ensure that each source track is copied to its equivalent on the DVD stream. Dolby specifies that the sequence for the channels is Left, Right, Low Frequency Effect (LFE), Left Surround and Right Surround. However, it cannot be assumed that all production houses follow that standard. As explained in Chapter 3, a problem arises when converting from NTSC to PAL, or transforming from cine film to PAL. In either case, the audio must match the change in the video length, with a 4 per cent speed up. The pitch must also be changed to ensure the proper pitch is restored. If the audio is from an audio CD, the sound must be upsampled from 44.1 kHz to 48 kHz for DVD-video. DVD-audio supports a sampling rate of 44.1 kHz and no upsampling is necessary, unless of course a Digital Dolby audio version is also included.

Sub-picture assets

The use of sub-pictures is widespread in DVD applications. These can emphasize or de-emphasize regions on the screen, tint a video, and provide highlights for selected menu choices. They are most useful for subtitles, and because they can be forced to appear under program control they are especially useful for multilingual DVD discs. They can also be used to create a limited animation overlaying the video. A DVD player can display up to four different colours at a time overlayed on the main image. The areas or regions of the screen in which each of these colours are to be displayed are defined by a mask that is prepared separately. The colours used are not those that will be displayed on the screen; the mask colours are defined in the DVD specification, with white for background areas, black for pattern areas, and red and blue for emphasis. During playback, the player uses the mask information (specified areas and associated colour) to determine where the background, pattern or emphasis is to fall on the screen. The actual colour of the overlayed areas is drawn from a 16-colour pallet. The transparency of each colour is defined at the authoring stage; for example, a background sub-picture has a transparency of 0 per cent to allow underlying images to come through.

Menus and graphics

Preparation of menus and associated graphics is normally carried out by computer-based programs. For this reason, the distinction between a DVD display and a computer-based display in terms of colour palettes and image dimensions is important. The range of colours supported by PAL and NTSC television systems is different from that supported by computers, and care must be taken to ensure that colours generated on a computer-based display will translate adequately when displayed on a television screen. As for image size, computer displays use '*square*' pixels while television displays use '*rectangular*' pixels – thus if no action is taken, a graphic display designed on a computer-based machine will be elongated when displayed on a television screen. To prevent this, a video graphic at 768 × 576 pixels should be created to be scaled down to 720 × 576 pixels before encoding (Figure 12.2). For NTSC, the graphic should be designed at 720 × 540 and scaled down to 720 × 480 pixels. For 16 : 9 anamorphic video, the equivalent figures are 960 × 576 scaled down to 720 × 576 pixels for PAL, or 720 ×

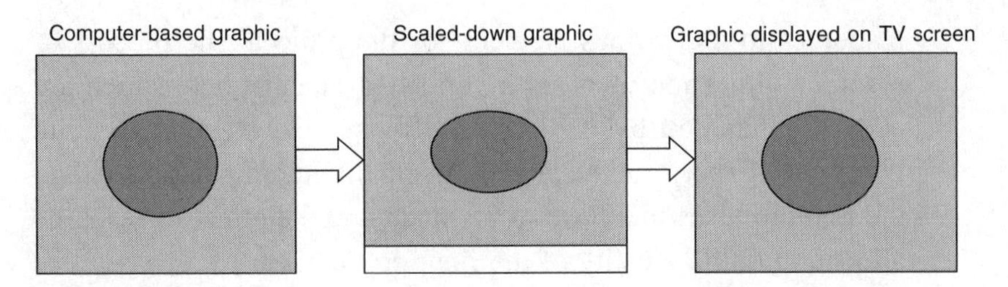

Figure 12.2

480 pixels for NTSC. The scaled-down graphics will look vertically compressed on a computer-based display, but appear at the correct aspect ratio when displayed on a widescreen television.

A further consideration when displaying computer-designed graphics on a television screen is the effect of over-scanning by the television system. The electron beam of a television cathode ray tube is made to over-scan the screen by as much as 10 per cent, which could result in a graphic that is intended to appear at or near to an edge of the picture not being displayed in part or in whole. To avoid this, 'safe areas' are defined to provide a boundary for the region of the screen in which text and other vital information should be kept (see Figure 12.3).

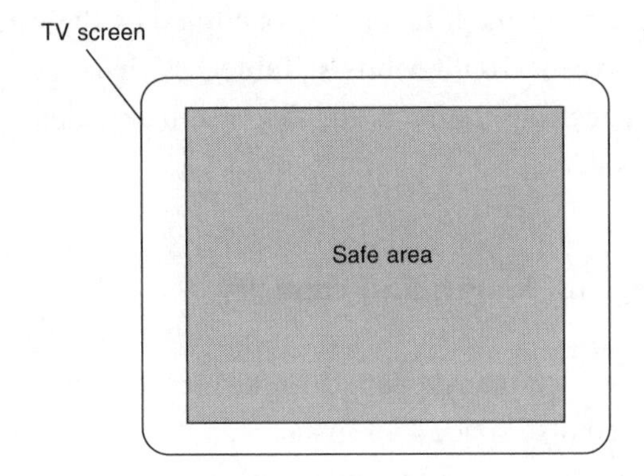

Figure 12.3

Resource allocation

Once the list of assets has been completed and the resource requirements are known, the next step is the allocation of these resources to the various elements of the project. The process of resource allocation involves a variety of calculations, which are normally handled by dedicated software. However, it is quite instructive to perform the process manually in order to understand the compromises that have to be made and the effect of resource allocation on the quality of the final product.

A DVD disc has two resources, capacity and bit rate (also known as *bandwidth*), both of which have strict limitations that cannot be exceeded. Although the capacity of a DVD disc is far greater than that of the traditional CD-ROM, it is still limited to 4.7, 8.54, 9.4, 13.24 or 17 GB, depending on the type of disc used – i.e. single-sided, single-sided dual layer, double-sided, or double-sided dual layer respectively. Added to capacity, standard DVD players have a maximum track bit rate of 10.08 Mbits/s. Within this maximum bit rate, each individual element (video, audio, sub-picture, etc.) has its own individual bit rate that must not be exceeded, namely 9.8 Mbits/s for the video element, 6.144 Mbits/s for an eight-channel linear PCM audio, 0.448 Mbits/s for 5.1 Dolby Digital, 0.384 Mbits/s for MPEG-1 stereo, 0.912 Mbits/s for MPEG-2 surround sound audio, and 3.36 Mbits/s for sub-picture streams. At any one time, the total bit rate occupied by the various elements must not exceed 10.08 Mbits/s. Table 12.1 lists the typical bit rate or disc space requirements for the various elements that make up a DVD title.

Bit rate, program length and capacity

For any given bit rate, the length of the recorded program determines the required disc space. For instance, a data rate of 10 Mbits/s and a program length of 30 minutes will require a disc capacity of

$$30 \text{ min} \times 60 \text{ s per min} \times 10 \text{ Mbits/s} = 18\,000 \text{ Mbits}$$

$$= \frac{18\,000}{1000} \text{ GB}$$

Table 12.1 Bit rate for various DVD elements

DVD	*Bit rate Mbits/s*	*Size Mbits*
Audio track Dolby Digital 5.1	0.384 or 0.448	
Audio track Dolby Digital stereo	0.192	
Audio track MPEG-1	0.192–0.384	
Audio track MPEG-2	0.384–0.912	
Audio track linear PCM (no data reduction)	6.114	
Audio track linear PCM (with MLP data reduction)	3.668	
Sub-picture per track	0.01	
Subtitle	0.04	
Still image		1.00
MPEG-2 video	Up to 9.8	

$$= \frac{18}{8} \text{ GB}$$

$$= 2.25 \text{ GB}$$

If the bit rate were reduced to 5 Mbits/s, then the disc space required for a 30-minute program would be:

$$\frac{30 \times 60 \times 5}{8 \times 1000} = 1.125 \text{ GB}$$

If the program length were to double to 60 minutes, then for a bit rate of 10 Mbits/s, the disc space required would be:

$$\frac{60 \times 60 \times 10}{8 \times 1000} = 4.5 \text{ GB}$$

and so on for different bit rates and different program lengths. Figure 12.4 shows a graphical representation of the relationship between bit rate and disc space for different program lengths, and how they fit into the standard disc sizes.

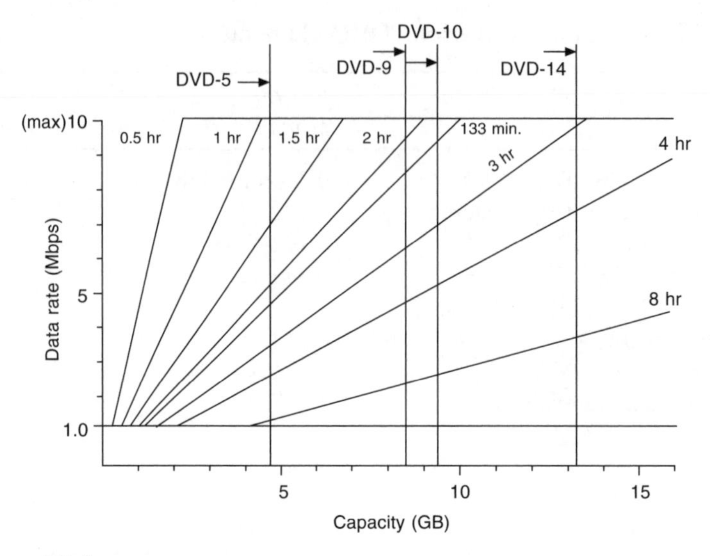

Figure 12.4
Data rate versus capacity

Bit budgeting

Bit allocation normally starts with the capacity of the disc, the length of the film being recorded, and the audio configuration to be used. Once these have been ascertained the bit rate may then be allocated to the various elements, such as audio, video, text, sub-pictures and interactivity. This process is known as bit budgeting.

Bit budgeting normally starts with the audio configuration. This is because with audio, the bit rate requirements follow directly from the configuration in terms of the number of channels, sampling rate, and quantization. The result is a more-or-less constant fixed bit rate that cannot be varied. With video, MPEG-2 allows a low target bit rate to be assigned and forces the encoder to cope with it the best it can. A low video bit rate will of course result in poorer quality picture resolution, which must be avoided if at all possible. However, compromises may have to be made if all the elements of the planned structure are to be met. Having determined the bandwidth or bit rate allocation for the audio element of the film, the remainder may be used for the video content. Although in the majority of cases variable bit rate is employed for the encoding of video, sometimes VBR does not result in improvement in the quality of the picture and thus it need not be employed. Instead, the less complex and more straightforward constant bit rate (CBR) is used. The decision regarding whether to use VBR or CBR is taken as part of the bit budgeting process.

In determining the available disc space, a proportion of the space (usually 4 per cent) is normally reserved for '*overheads*' for such things as menu graphics, navigation information, or a simple safety cushion. Thus a DVD-5 (single-side, single-layer) with a capacity of 4.7 GB can only accommodate $0.96 \times 4.7 = 4.512$ GB. A DVD-9 disc (single-side, dual-layer) with a capacity of 8.54 GB has an available or adjusted disc space of $0.96 \times 8.54 = 8.1984$ GB. Similarly, DVD-14 (double-sided with one single layer and one dual layer) with a capacity of 13.24 GB has an available capacity of $0.96 \times 13.24 = 12.71$ GB and so on, as listed in Table 12.2.

Table 12.2 Available disc space for different sized DVD discs

Disc type	*Total capacity*		*Overheads*		*Available capacity*	
	GB	*Mbits*	*GB*	*Mbits*	*GB*	*Mbits*
DVD-5	4.7	37 600	1.88	1504	4.512	36 096
DVD-9	8.54	68 320	3.416	2733	8.1984	65 587
DVD-10	9.4	75 200	3.76	3008	9.024	72 192
DVD-14	13.24	105 920	5.296	4237	12.71	101 683
DVD-18	17.08	136 640	6.832	5466	16.397	131 174

Bit budgeting examples

Example 1

A single video title is to be produced, 120 minutes long, using Dolby Digital (AC3) 5.1 with a facility for six-track subtitles (i.e. subtitles in six different languages). The title is to be recorded on a DVD-5.

The starting point is to calculate the available space on the disc that may be used for recording the program.

$$
\begin{aligned}
\text{DVD-5 capacity} &= 4.7 \text{ GB} \\
&= 4.7 \times 10^9 \text{ bytes} \\
&= 4.7 \times 10^9 \times 8 \text{ bits} \\
&= 37.6 \times 10^9 \text{ bits} \\
&= 7600 \times 10^6 \text{ bits} \\
&= \mathbf{37\,600} \text{ Mbits}
\end{aligned}
$$

With 4 per cent of disc space reserved for overheads:

$$
\begin{aligned}
\text{Available disc space} &= 37\,600 \times 0.96 \\
&= \mathbf{36\,096} \text{ Mbits}
\end{aligned}
$$

Having determined the amount of disc space available for recording the video and related audio and subtitle data, the available resources – namely the disc space and bit rate – can be apportioned to the elements of the title. The title is a simple video program with AC3 sound and multi-language subtitles, and it therefore has three elements; audio, subtitles and video.

Starting with the audio stream, the Dolby Digital 5.1 format may be implemented using a bit rate from 0.384 to 0.448 Mbits/s. If 0.448 Mbits/s is selected, the disc space occupied by Dolby Digital 5.1 surround sound can be calculated to be:

$$
\begin{aligned}
\text{Audio bit rate} \times \text{length of program} &= 0.448 \times 120 \times 60 \\
&= 3\,226 \text{ Mbits}
\end{aligned}
$$

With each subtitle track requiring 0.04 Mbits/s, the total bit rate that has to be allocated for this element is:

$$0.04 \times 6 = 0.24 \text{ Mbits/s}$$

For the total length of the video, the total disc space that must be allocated to the subtitles element is:

$$0.24 \times 120 \times 60 = 1724 \text{ Mbits}$$

Therefore:

Available disc capacity – space taken up by the audio element – space taken by the subtitle element

$$= 36\,096 - 3226 - 1724 = \mathbf{31\,146} \text{ Mbits}$$

This remaining disc space may now be used to record the video component of the program, giving an average video bit rate of:

$$\frac{\text{Disc space for video}}{\text{Time of program}} = \frac{31\,146}{120 \times 60} = \mathbf{4.326} \text{ Mbits/s}$$

Although in general it is more efficient to use a variable bit rate for the video element, this is only so if the average bit rate is below the maximum available bit rate. The maximum available bit rate for the video element may be calculated by adding the bit rates of all the non-video elements that occur simultaneously with the video element and taking that total away from the maximum available bit rate, namely 10.08 Mbits/s. In this example there are two non-video elements in the DVD program that are played back simultaneously with video. Thus the total simultaneous bit rate is:

$$\text{Audio bit rate} + \text{subtitle bit rate} = 0.448 + 0.24 \text{ Mbits/s}$$
$$= 0.688 \text{ Mbits/s}$$

The maximum available bit rate for the video element is:

Maximum track bit rate – simultaneous non-video bit rate

$$= 10.08 - 0.688$$
$$= 9.392 \text{ Mbits/s}$$

Since the average video average bit rate (4.326 Mbits/s) is smaller than the maximum available bit rate (9.329 Mbits/s), it is possible to optimize the image quality by using variable bit rate (VBR) encoding with a target bit rate of 4.326 Mbits/s and a peak rate of 9.392 Mbits/s. This means that for fast-moving and relatively more detailed scenes the maximum available video bit rate (namely 9.392 Mbits/s) may be used, while easy-to-encode images may be allocated fewer bits using a bit rate lower than the 4.565 Mbits/s average, resulting in an overall enhanced quality video. If the film was shorter and/or a more bit rate-hungry audio configuration was used, the average bit rate could be so high that a constant bit rate (CBR) would be more efficient (see Example 2).

Example 2

In this example, the same three elements as in Example 1 are present – i.e. a video title with Dolby Digital (AC3) 5.1 surround sound and six-track subtitles – but the length of the film is now reduced to 50 minutes. The title is to be recorded on the same disc type (DVD-5).

As above, allowing for 4 per cent overheads, the adjusted capacity of the DVD-5 disc is:

$$0.96 \times 4.7 \times 1000 \times 8 = 36\,096 \text{ Mbits}$$

The 5.1 Dolby Digital audio element at bit rate of 0.448 Mbits/s will occupy a disc space of:

$$0.448 \times 50 \times 60 = 1344 \text{ Mbits}$$

The six-track subtitle will occupy:

$$0.04 \times 6 \times 50 \times 60 = 720 \text{ Mbits}$$

Therefore:

Available disc space – disc space taken up by the non-video elements

$$= 36\,096 - 1344 - 720 = \mathbf{34\,032} \text{ Mbits}$$

This remaining disc space may now be used to record the video component of the program, giving an average video bit rate of:

$$\frac{\text{Disc space for video}}{\text{Time of program}} = \frac{34\,032}{50 \times 60} = \mathbf{11.344} \text{ Mbits/s}$$

The maximum available bit rate is the same as that calculated in the previous example:

$$\begin{aligned} \text{Maximum available bit rate} &= \text{maximum track bit rate} \\ &\quad - \text{simultaneous non-video bit rate} \\ &= 10.08 - 0.448 - 0.24 \\ &= \mathbf{9.392} \text{ Mbits/s} \end{aligned}$$

In this example, the average video bit rate (11.344 Mbits/s) is higher than the maximum available bit rate (9.392 Mbits/s). Here, the actual encoding bit rate used would be the lower figure – i.e. the maximum available bit rate of 9.392 Mbits/s. The higher figure (i.e. that calculated to be the average bit rate) could not be used. It follows that in such cases there would be no advantage in using a variable bit rate rather than the more straightforward constant bit rate.

Example 3

A DVD-9 audio title is to be produced, designed for both the UK and French markets. The title is made up of two parts; a 120-minute live video concert recording with Dolby Digital sound, and a 60-minute music section containing 12 tracks of high-quality PCM surround sound. The video concert is to be accompanied by subtitles in two languages, and the music section by still images and subtitle text in two languages.

The elements of the title are therefore:

Audio section:			
PCM	Six tracks	60 min	24-bit/96 kHz
Still images			1.0 Mbits
Subtitles	Two tracks (English and French)	60 min	0.04 Mbits/s per track
Video section:			
5.1 Dolby Digital		120 min	0.448 Mbits/s
Video		120 min	
Subtitles	Two tracks (English and French)	60 min	0.04 Mbits/s per track

The audio section:

The bit rate for six tracks of PCM audio using a sampling frequency of 96 kHz and a word length of 24 bits is:

$$96 \text{ kHz} \times 24 \text{ bits} \times 6 \text{ tracks} = 13\ 824 \text{ kbits/s}$$

$$= \frac{13\ 824}{1000} \text{ Mbits/s}$$

$$= \mathbf{13.824} \text{ Mbits/s}$$

This bit rate is far in excess of the 10.08 Mbits/s maximum bit rate of a DVD player, and therefore cannot be used. The only way to retain the high resolution PCM sound is to use *meridian lossless packing* (*MLP*) for audio data compression. MLP compresses data bit by bit, removing redundant data without any loss to quality, and achieves a compression ratio of 2 : 1. However, in practice a data compression ratio of 0.6 is normally used. Using this figure, the bit rate for the audio PCM element may now be reduced to a more practical level, namely:

$$0.6 \times 13.824 = 8.294 \text{ Mbits/s}$$

The disc space that must be allocated to the audio PCM element is:

Audio bit rate × length of music part

= 8.294 × 60 min × 60 s per min

= **29 858** Mbits

Still images:

Assuming one still image per track, with each image occupying 1.0 Mbits, the total disc space for the still images is:

1 Mbit × 6 = **6** Mbits

Subtitles:

With subtitles requiring a bit rate of 0.04 Mbits/s per track or language, the total bit rate for the subtitle element is:

2 tracks × 0.04 Mbits/s = **0.08** Mbits/s

Consequently, the disc space required for this element is:

Bit rate × time of music section = 0.08 × 60 × 60 = **288** Mbits

Therefore:

Total disc space allocated to the music section

= disc space for PCM audio + disc space for still images + disc space for subtitles

= 29 858 + 6 + 288 = **30 152** Mbits

The video section:

The video concert consists of three elements:

1. 120 minutes of Dolby Digital audio.
2. 120 minutes of video.
3. 120 minutes of the two-track subtitle element.

The video-related Dolby Digital element requires a bit rate of 0.448 Mbits/s, thus occupying a disc space of:

$$0.448 \times 120 \times 60 = \mathbf{3226}\ \text{Mbits}$$

The subtitles element requires a bit rate of 0.04 Mbits/s per track or language, giving a total bit rate for the subtitle element of:

$$2 \times 0.04 = \mathbf{0.08}\ \text{Mbits/s}$$

Consequently:

Disc space required for the subtitle element = $0.08 \times 120 \times 60$

$= \mathbf{576}$ Mbits

The total non-video element of the concert is 3226 + 576 = 3802 Mbits. It is now possible to calculate the total disc space used by all the non-video elements, which is:

Disc space for the music section + disc space for the non-video elements of the video concert

= 30 152 + 3802 = **33 954** Mbits

The total capacity of the DVD-9 is 8.54 GB = 68 320 Mbits. Therefore, allowing for 4 per cent overheads:

Available disc capacity = $0.96 \times 68\,320$ = **65 587** Mbits

With 33 378 Mbits being used for the non-video elements:

Remaining disc space available for the video elements

= available disc capacity – disc space for the non-video elements

= 68 320 – 33 954 = **34 366** Mbits

Therefore:

$$\text{Average video bit rate} = \frac{\text{video disc space}}{\text{length of video element}}$$

$$= \frac{34\,366}{120 \times 60}$$

$$= \mathbf{4.77}\ \text{Mbits/s}$$

The maximum or peak video bit rate is:

10.08 – bit rate for all elements simultaneous with the video element.

The elements that are played back simultaneously with the video element are:

- 120 min Dolby Digital, bit rate 0.448 Mbits/s
- 120 min two-track subtitles, bit rate 0.08 Mbits/s.

It follows that:

Maximum video bit rate = 10.08 – 0.448 – 0.08 = 9.512 Mbits/s

The video encoder should then be set for a variable bit rate having an average of 4.77 Mbits/s and a peak of 9.512 Mbits/s.

Authoring

Once all the assets have been prepared and the available resources allocated, the various elements have to be brought together in the authoring process. Authoring comes is a media integration process, where the individual elements are combined in a unified whole with navigational pathways linking the various parts. Up until the introduction of DVD, audio- and video-recorded programs did not need authoring, or if they did it was very limited – such as indexing for VHS. Apart from simple indexing, VHS production does not require any program logic to be introduced into the video or audio content. Indeed there are no such facilities with such productions. With audio-CD, the format supports some basic navigation to provide the listener with select modes that can select a track or vary the playback order.

With DVD-based formats, authoring is far more complex. DVD authoring organizes the presentation data (video, audio, graphics) into video objects (VOBs) and the navigation data into program chains (PGCs), and integrates the two parts to form a single product.

The first step in authoring is to prepare what is known as a *storyboard*, which would initially have been worked out in the planning stage. The next step is to import and assemble all of the prepared source files – video, audio, menu, graphic, subtitles and sub-pictures. Video and audio files are more likely to be in the form needed for DVD, as they will have been prepared and encoded at the previous stage. Graphic files, on the other hand, are normally in bitmap format, and will need to be converted to MPEG stills. The advanced authoring software environment allows for file import, including bitmap conversion, to be handled as a simple drag-and-drop activity. Once the files have been imported, they are linked together to construct video and audio objects. These basic objects appear on a timeline against which additional streams such as audio, subtitles and stills may be laid and synchronized.

The construction of VOBs is followed by the creation of the menu. Having imported the stills for the menu, the areas to be highlighted are defined with specific colours and 'hot spots' are created for buttons. This part of the process may be extremely complex and highly labour-intensive. However, it is an essential aspect of the authoring process, providing a graphical interface for sub-picture assembly that allows fast and intuitive control of various features by the user.

Once the objects have been constructed and assembled and the menus laid out, program chains (PGCs) are created to provide title interactivity. DVD supports complex and sophisticated facilities for jumps, links and branching, and other navigational utilities for such applications as training and games. Pre- and post-commands for various PGCs are inserted to provide such facilities as multiple angle and parental control.

DVD authoring includes a utility for defining a relationship between Internet HTML pages and the video material. URLs and browser frames may be directly linked with menu buttons, as well as specific points in the video stream.

In any authoring software environment, a facility for instant preview and proofing is essential. The preview facility should permit the program to be viewed and heard at full resolution, so that interactivity may be fully checked.

Formatting

The final stage of the production of a DVD title is the formatting of the material prepared by the authoring process to create the final disc image necessary for the construction of a master disc. Formatting is entirely determined by decisions and definitions made in the authoring stage, and once initiated requires no further input. It involves multiplexing the various assets and their objects, adding navigation and control information according to the DVD specification. The program chains are incorporated into the video manager information (VMGI); the audio manager information (AMGI) or the video title set information (VTSI) areas as appropriate. In addition, the volume information for titles with multiple volumes is generated. The end result of the formatting process is to produce a new set of files that comply with the format specification for DVD-video and DVD-audio.

CHAPTER 13

DVD DRIVES

Introduction

A personal computer is a general purpose microprocessor-based system that carries out a variety of function and operations, including data manipulation, word processing, and reading and recording of optical storage devices such as CD ROMs and DVD discs. A basic personal computer consists of a *system box* and a number of peripheral devices, such as a keyboard and a mouse (Figure 13.1).

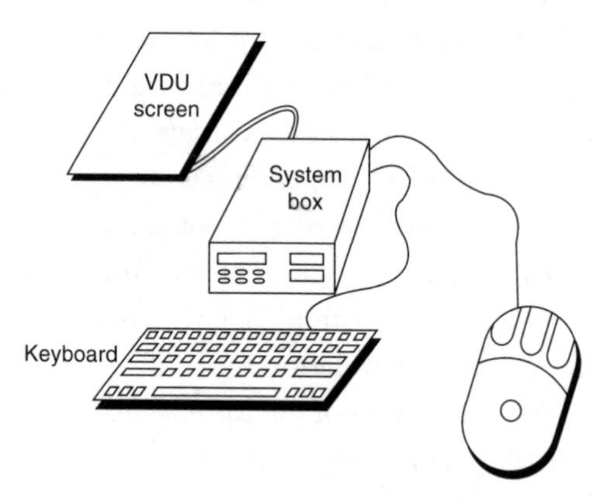

Figure 13.1

The system box has two elements: hardware and software. The hardware elements (Figure 13.2) include the power supply, the motherboard, the floppy and hard disk drives, and various ports such as parallel and serial ports and USB (universal serial bus) ports. The software consists of the various programs and routines that are required by the microprocessor and other support chips in order to function. The software may be in the form of a programmed ROM or EPROM chip, or it may be stored on a floppy disk, a hard disk or CD/DVD discs.

The main hardware element of the system box is the motherboard, which is shown in simplified form in Figure 13.3. The central processing unit or microprocessor carries out the programming and control of the whole system. In this task, it is aided by a set of dedicated support chips to carry out such activities as clock generation and IRQ and DMA control. The start-up routine, known as the BIOS (basic input output system) routine, is stored in a ROM memory chip known as the ROM BIOS. System memory is

Figure 13.2

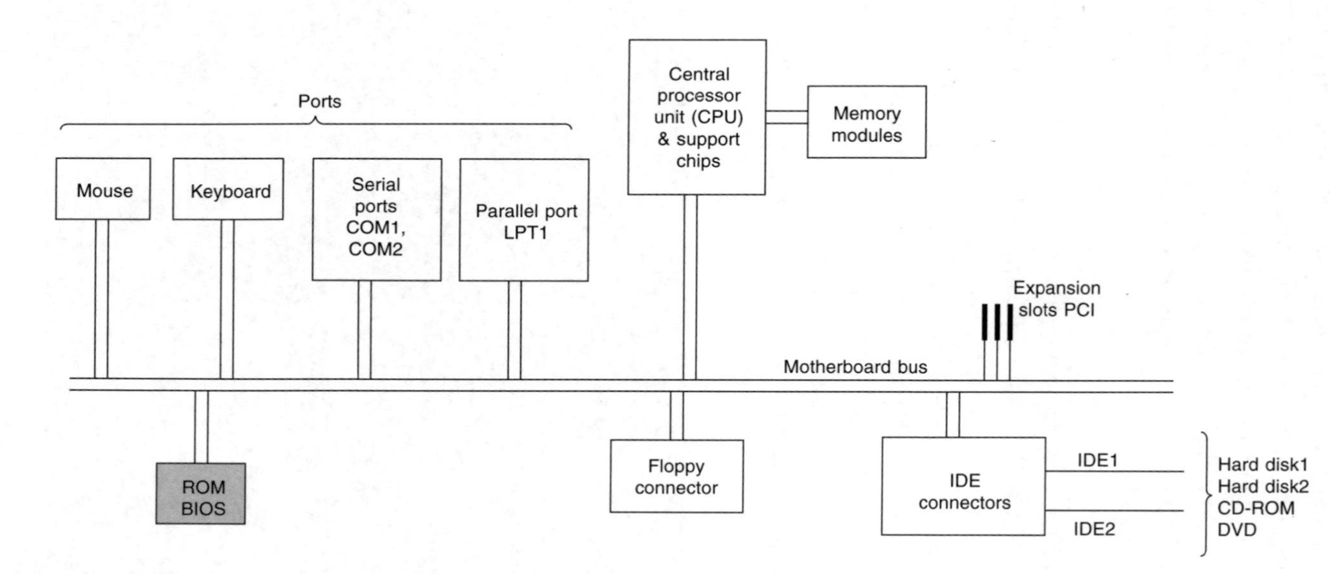
Ports
Mouse
Keyboard
Serial ports COM1, COM2
Parallel port LPT1
Central processor unit (CPU) & support chips
Memory modules
Expansion slots PCI
Motherboard bus
ROM BIOS
Floppy connector
IDE connectors
IDE1
IDE2
Hard disk1
Hard disk2
CD-ROM
DVD

Figure 13.3

provided by a number of DRAM or SDRAM modules, normally dual in-line memory modules (DIMMs).

The speed of operation of the motherboard is determined by the speed of the processor, and the size and speed of the system memory. A 1-GHz processor speed is not unusual, and a memory of 128 MB is the norm for most applications. While system memory is of the DRAM or SDRAM variety, with a relatively slow access speed of 30–60 μs, faster but more expensive SRAM chips may be used to speed up operations. SRAM chips are used as *cache* memory where the more frequently used instructions and data are written for fast access by the CPU. Although small in size (typically 128–512 KB), cache memory can produce a vast improvement in the computing power of the system. The motherboard provides several ports, including two serial ports and one parallel port, for a number of devices such as a mouse, a keyboard, and video and audio outputs. A parallel port may be used for a printer or a scanner, and serial ports may be used for several devices, including an external modem.

A typical motherboard (Figure 13.4) provides a number of connectors for storage devices, such as floppy and hard disk drives and CD-ROM and DVD disc drives. Motherboards support one 34-way floppy disk connector and two 40-way *IDE* (*integrated device electronics*) connectors. The IDE connectors are referred to as a primary IDE1 and a secondary IDE2, and are used for hard disk drives and CD/DVD disc drives. The floppy disk drive connector (FDC1) can support two floppy disk drives connected to the same cable as *master* and *slave*. Similarly, each of the two IDE connectors can support two drives connected as master and slave (see Figure 13.5). Pin 1 is identified by a stripe (normally red, but sometimes blue) running along one side of the ribbon cable, and when connected to the device, pin 1 is nearest to the power supply connector. The device is set to master or slave by jumper settings, and these settings are indicated on the device itself. An example of a jumper setting is shown in Table 13.1.

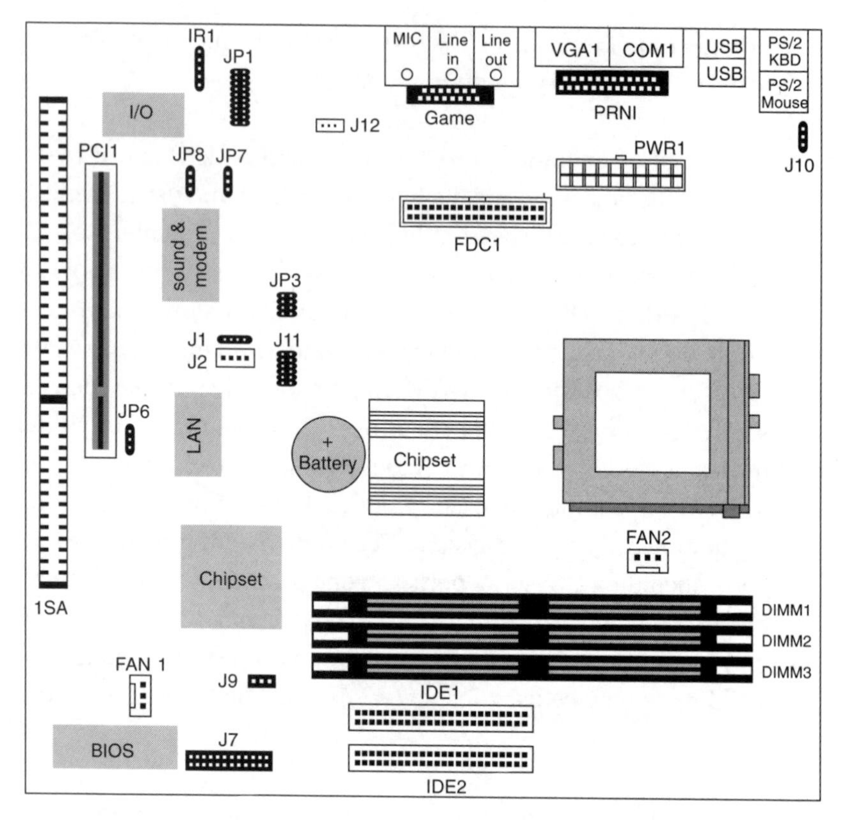

Figure 13.4

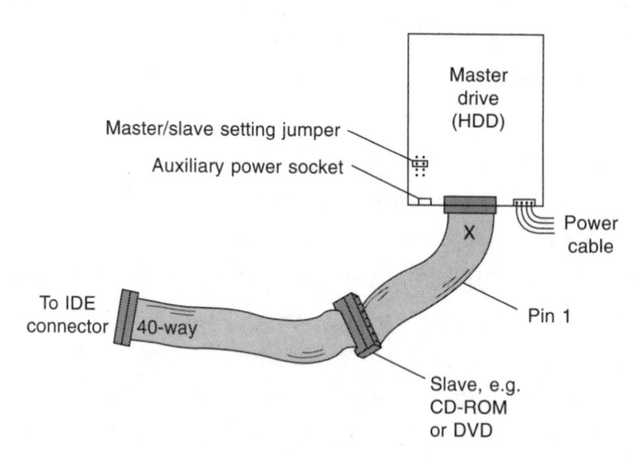

Figure 13.5

Table 13.1 Example of a jumper setting

Name	*Function*	*CSEL*	*Slave*	*Master*
MA (Master)	Drive set as Master	○ ○	○ ○	▮
SL (Slave)	Drive set as Slave	○ ○	▮	○ ○
CS (Cable Select)	Drive mode set by CSEL on the host IDE interface	▮	○ ○	○ ○

If you use CSEL setting, the MASTER/SLAVE setting will be made automatically, depending on the hardware configuration. For details, refer to the manual of your computer.

Another way of connecting devices such as a hard disk drive or a DVD drive is via the SCSI. Unlike IDE connections, which require external controllers in the form of either a card or (as is the case in modern PCs) integrated on the motherboard, SCSI devices have most of their controller electronics in the drive itself. SCSI devices, such as a scanner, hard disk drive or a DVD drive, are connected to a SCSI host adaptor card inserted in one of the slots on the motherboard using a 50-way cable. More than one device may be connected in a chain (see Figure 13.6). Each device, including the controller card (also known as the host card) is identified by an individual ID number or address assigned to it when it is connected to the bus. There are seven ID numbers in total, with the SCSI controller card itself usually assigned ID number 7. ID numbers

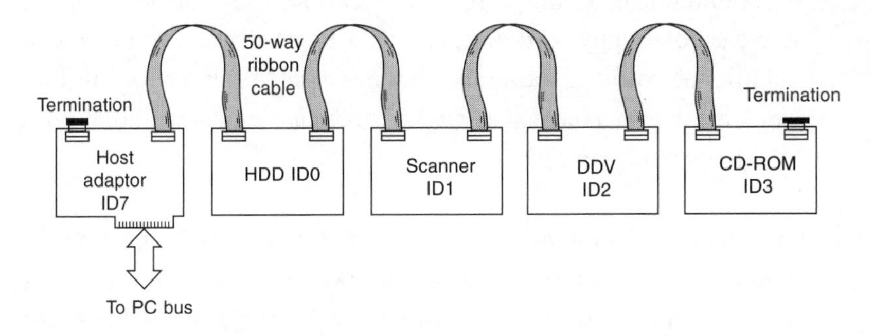

Figure 13.6

are set using hardware control on the SCSI device itself, usually jumper or dipswitch settings (see Figure 13.7).

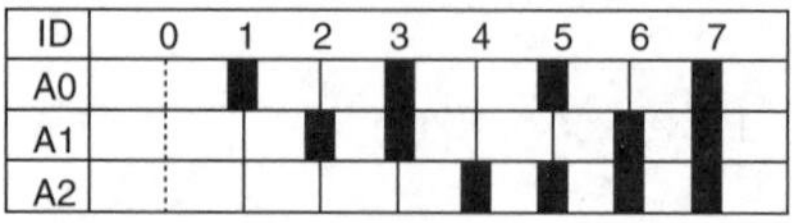

ID	0	1	2	3	4	5	6	7
A0								
A1								
A2								

Figure 13.7
SCSI ID 54 jumper setting

Devices such as a modems or a sound card may be added to the system by inserting them into one of the expansion slots. Although there are different types of expansion slots, the most popular are the old ISA (industry standard architecture) and PCI (peripheral component interconnect) slots. The motherboard in Figure 13.4 shows one PCI slot (PCI1) and one ISA slot.

Operating system

When a PC is switched on the processor looks for and runs the start-up routine, which resides in the ROM BIOS chip. One of the tasks of the start-up routine is to load generic drivers and appropriate software for a number of basic peripheral devices such as the keyboard, video monitor, floppy and hard disk drives, and the CD ROM drives. The purpose of the drivers is to enable the processor to communicate with, program and control these devices and, in the case of floppy/hard disks and CD ROM drivers, to download additional software programs if necessary. Without any further software the machine will halt, with a message indicating the absence of an operating system.

An operating system is essential if the user is to be able to communicate with the computer via a keyboard or a mouse, and if the computer is to run applications such as word processing or Internet browsing. The operating system is normally found on the hard disk and, if present, will be picked up automatically by the processor as part of the BIOS routine. The operating system will load drivers and the relevant software for every device that has been installed, resulting in a fully functioning personal computer. The process of loading BIOS and the operating system is known as the boot-up routine of the computer.

There are several operating systems, including DOS (disk operating system), UNIX, and Microsoft's series of Windows operating systems. By far the most popular are Windows 95/98 and Windows 2000. For the purposes of this section, the main difference between the various operating systems is the way in which they recognize and instal devices such as CD or DVD drives. Windows 2000, for instance, is a fully '*plug-n-play*' (*PnP*) operating system – i.e. it will recognize and instal external devices automatically. For other operating systems, such as DOS, installing external devices involves manually editing a number of files, allocating resources, and loading the necessary drivers and programs.

Allocating resources

When a device is connected to a computer it must be allocated a number of resources, such as an interrupt request, a port address and a DMA channel. Without these resources, the device cannot be identified by the CPU. The main resource that a device must be allocated is an *interrupt request* (*IRQ*). A PC has a total of 16 IRQs (IRQ0–IRQ15), of which only 15 may be used. A number of these IRQs are assigned for specific functions, such as IRQ0 for the system clock and IRQ14 for the first hard disk drive (HDD1), as shown in Table 13.2. Only five interrupt requests, namely IRQ5, IRQ10, IRQ11, IRQ12 and IRQ15, are available for allocation to additional devices such as a modem, a CD-ROM drive or a DVD drive. The allocation of IRQ may be carried out manually or, as is the case with most modern computers, automatically using plug-n-play routines. In either case, each device must be allocated its

Table 13.2 Allocation of interrupt requests in a computer

IRQ0	System timer
IRQ1	Keyboard controller
IRQ2	Cascade
IRQ3	Serial port COM2/COM4
IRQ4	(Shared) serial ports COM1/COM3
IRQ5	Available
IRQ6	Floppy disk controller
IRQ7	(Shared) parallel ports LPT1/LPT3
IRQ8	Real-time clock
IRQ9	Video unit
IRQ10	Available
IRQ11	Available
IRQ12	Available
IRQ13	Maths processor
IRQ14	Hard disk drive controller IDE1
IRQ15	Available

own individual IRQ number. If two devices are allocated the same interrupt request number, then *conflict* is said to occur and one or both devices will fail to function. Conflict is one of the most common malfunctions in personal computers.

Fault finding utilities

Windows supports a number of utilities that provide information about the system, including size of memory, type of processor, conflicts, and installed drivers to assist in diagnostics and faultfinding. There is, of course, the Help button, which provides tips on troubleshooting. The main troubleshooting utilities are System Information, the Advance Options Menu, and the Device Manager.

System information

System information is a very useful utility that provides detailed information regarding the operating system and hardware components of the computer. To open system information:

1. Click on ‘Start’
2. Select ‘Settings’
3. Select ‘Control Panel’ (Figure 13.8)
4. Double-click on ‘Administrative Tools’
5. Select ‘Computer Management’ (Figure 13.9)
6. Open ‘system information’, which will then give a number of options, including ‘Summary’ (Figure 13.10) and ‘Hardware Resources’
7. Open ‘Hardware Resources’ and select ‘IRQS’ to display a list of allocated interrupt requests (Figure 13.11). Other options are also available, such as ‘conflicts’ and ‘DMA’.

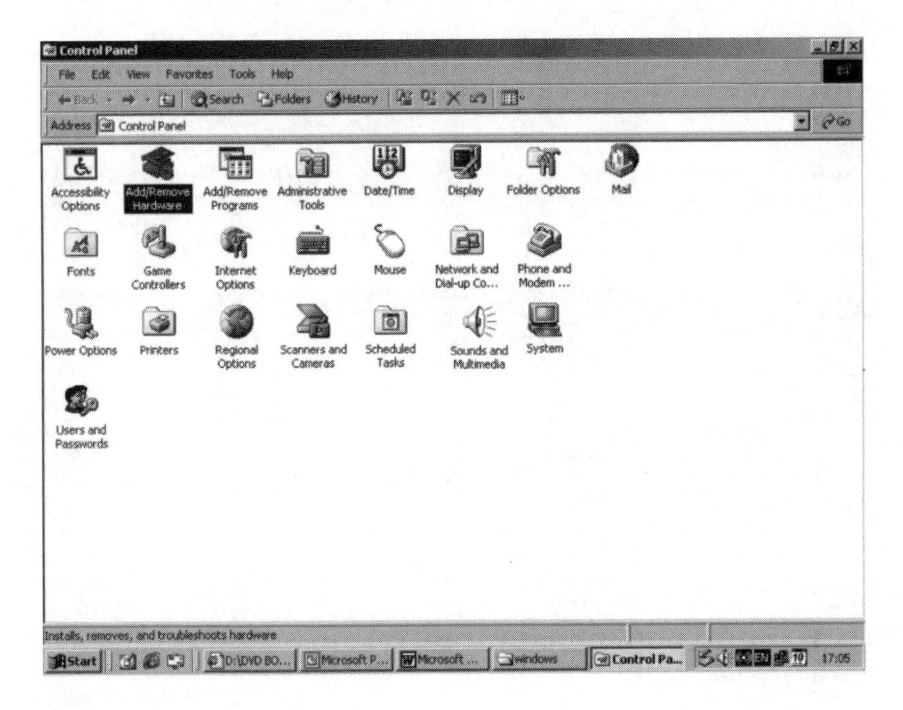

Figure 13.8
Control panel

The advanced options menu

The advanced options menu (Figure 13.12) is mainly used in cases of failure to boot up. It allows the computer to boot up with fewer

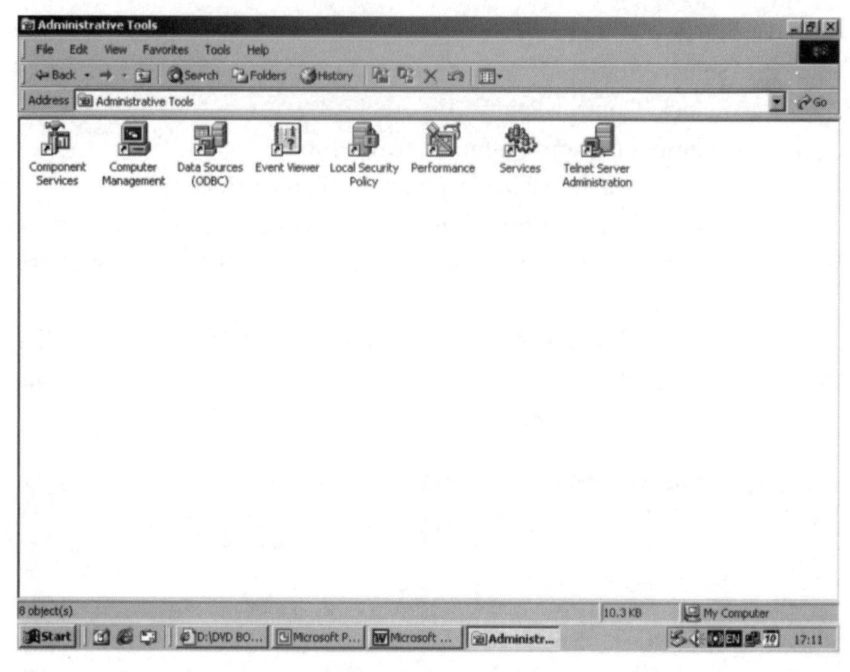

Figure 13.9

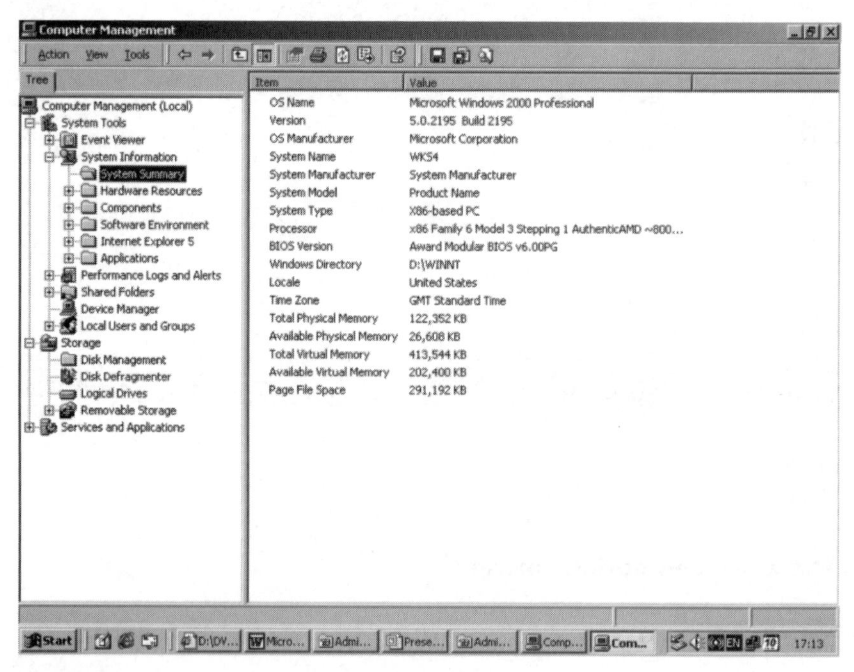

Figure 13.10

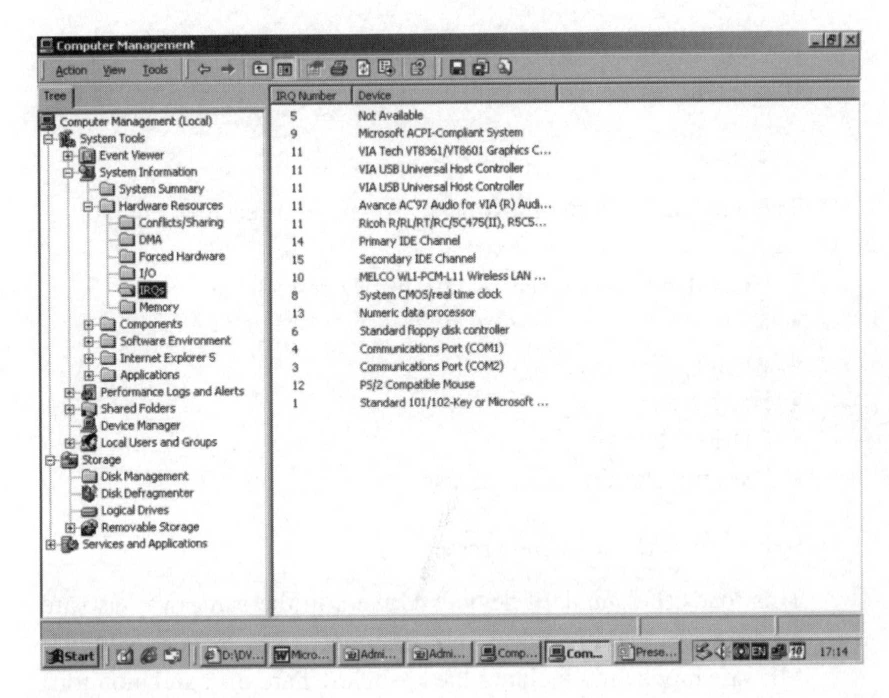

Figure 13.11

devices and programs, so that an examination of the hardware and the software can take place. The advanced options menu may be brought up by pressing F8 during the boot-up process. It provides a number of options that may be used in cases where normal

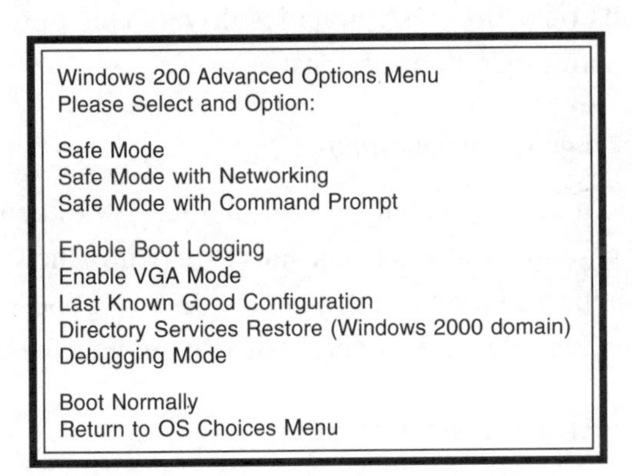
Windows 200 Advanced Options Menu
Please Select and Option:

Safe Mode
Safe Mode with Networking
Safe Mode with Command Prompt

Enable Boot Logging
Enable VGA Mode
Last Known Good Configuration
Directory Services Restore (Windows 2000 domain)
Debugging Mode

Boot Normally
Return to OS Choices Menu

Figure 13.12
Windows 2000 Advanced Options Menu

booting up is not possible or desirable, and these include the following.

Safe mode

This enables only the most basic drivers and services, including:

- VGA 4-bit colour, 640 × 480, 60-Hz refresh rate
- Mouse
- Keyboard
- Hard disk
- CD-ROM drive
- PS/2 mouse (not serial mouse).

Safe mode with command prompt

This loads the standard devices' drivers in the same way as safe mode, with a command-prompt interface. (The Windows 95/98/ME safe mode only includes the keyboard, hard disk and monitor.)

Safe mode with networking

This loads the safe mode devices with the necessary drivers and services to support networking.

Enable VGA mode

This loads only the VGA graphics driver. This option is enabled with all safe mode boot options.

Last known good configuration

When the machine is switched off, the Registry keeps a record of the configuration. If the machine fails to boot up, a new configuration may be bypassed by selecting the last known good configuration. Examples include cases where new drivers have been loaded.

Directory services restore mode

This restores a corrupt directory database. It is only for servers that are domain Directory Controllers.

Debugging mode

This initiates the debugging process to gather debugging data for future diagnostics. It is available only on Windows 2000 servers.

The device manager

The device manager is a utility introduced by Microsoft into Windows 95/98/ME/2000, but not NT. It provides a graphical representation of all the devices that have been configured. Figure 13.13 shows a typical device manager display. It may be accessed in the following way:

1. Select 'Start'
2. Select 'Settings'
3. Click on 'Control Panel'
4. Double-click on 'System' icon

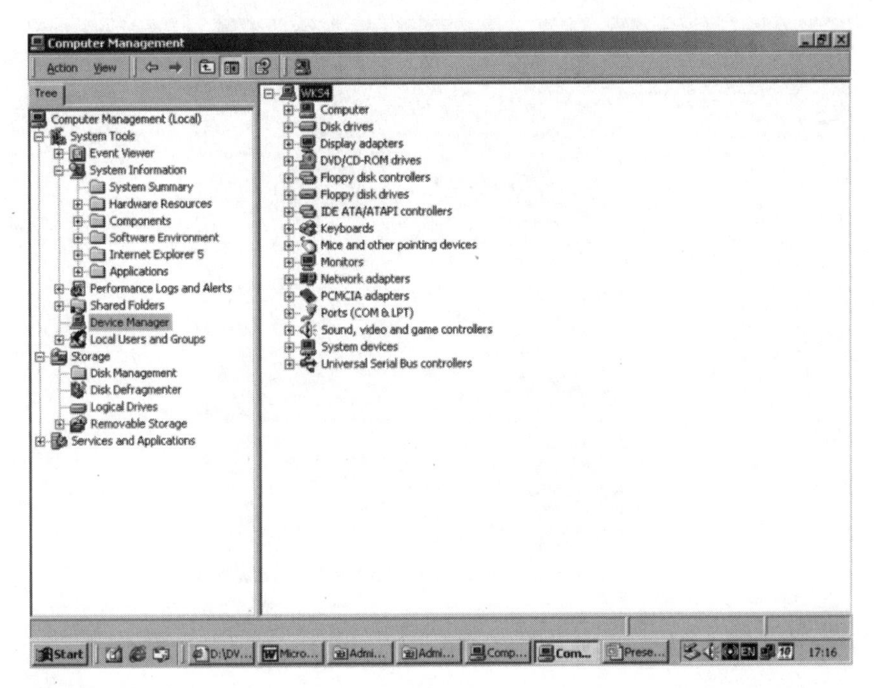

Figure 13.13

5. Click 'Hardware' on bar menu
6. Hit 'Device Manager' button.

The device manager lists all the hardware devices, their properties, drivers and the resources allocated to them. Problems are indicated by one of three red symbols:

!	Indicates that device is not present, is not installed, or there is a resource conflict
?	Indicates that not all drivers have been installed
X	Indicates sever resource conflict, with the offending item disabled

By double-clicking the offending symbol, the device properties, including its drivers and resource allocation (IRQ and port address), may be examined (Figure 13.14).

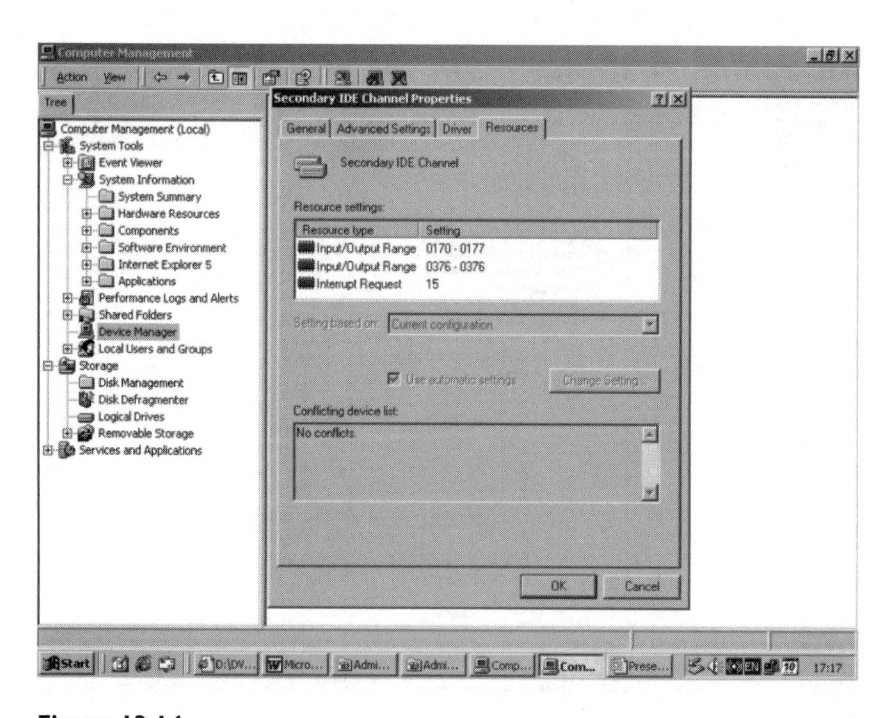

Figure 13.14

Installing a DVD drive

The first step in installing a DVD system is the installation of the physical drive. Installing a DVD drive is very similar to the installation of a CD ROM drive, and involves two stages: hardware and software installation. If the drive is to be used for video playback, an MPEG decoder card must also be installed. Some drives come with software for playing DVD movies, but unless the PC has a very fast CPU and large amounts of memory the playback may skip frames, thus causing jerky video reproduction. Other hardware requirements include an advanced graphic card with in-built assistance for MPEG-2 decoding. A good quality sound card is also required to do justice to the audio quality of DVD.

There are different types of DVD drives in terms of the way in which they connect to the computer. An IDE DVD drive uses one of the IDE connections provided by the motherboard, and this type is the most popular. However, other types are also available, including an SCSI, which uses a 50-way SCSI connection.

Installing an IDE DVD drive

An IDE DVD drive is almost indistinguishable from a CD-ROM drive. It has five different connections, which are provided at the rear of the device (see Figure 13.15). These are:

1. A 4-pin power connector
2. A 40-way IDE interface connector
3. A master/slave jumper setting
4. An analogue audio connector
5. A digital audio output connector.

There are three different ways in which a DVD drive may be connected to the motherboard. The motherboard provides two IDE connections, primary and secondary, and each IDE connector

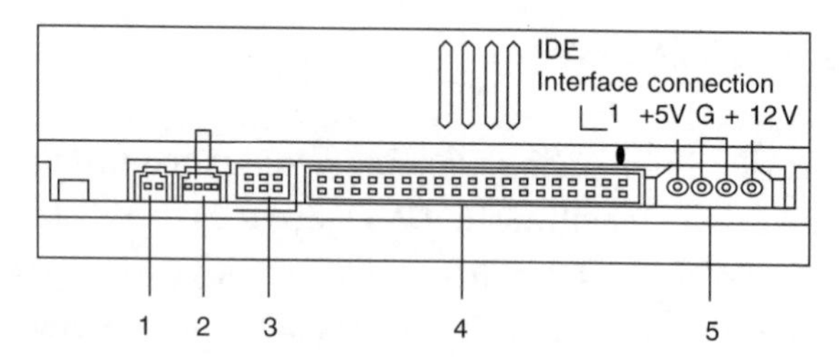

1. **Digital audio output connector**
2. **Analog audio output connector**
3. **Jumper connector**

   This jumper determines whether the drive is configured as a master or slave. Changing the master-slave configuration takes effect after power-on reset.
4. **IDE interface connector**

   Connect to the IDE (Integrated Device Electronics) Interface using a 40-pin flat IDE cable.

   Note: Do not connect or disconnect the cable when the power is on, as this could cause a short circuit and damage the system. Always turn the power OFF when connecting or disconnecting the cable.
5. **Power connector**

Figure 13.15

supports two devices, one as master and the other as slave. Normally, the boot hard disk is connected to the primary IDE1 as master. The DVD drive may then be connected to the IDE1 as slave (Figure 3.16a), or to the secondary IDE2 as master (Figure 13.16b). However, if another device is already connected to the IDE2 as master, then the DVD drive may be connected to IDE2 as slave (Figure 3.16c). DVD disc drives are shipped with jumper settings configured to master. If the drive is connected as a slave, the jumper setting must be changed accordingly.

Step-by-step installation procedure

Turn off your PC, unplug it, and remove the cover. Throughout any activity that involves inserting or removing cards or devices, precautions should be taken to avoid electrostatic discharge, which may cause damage to CMOS chips. For this reason you should be grounded, preferably with an antistatic wrist strap clipped to a grounded metal object.

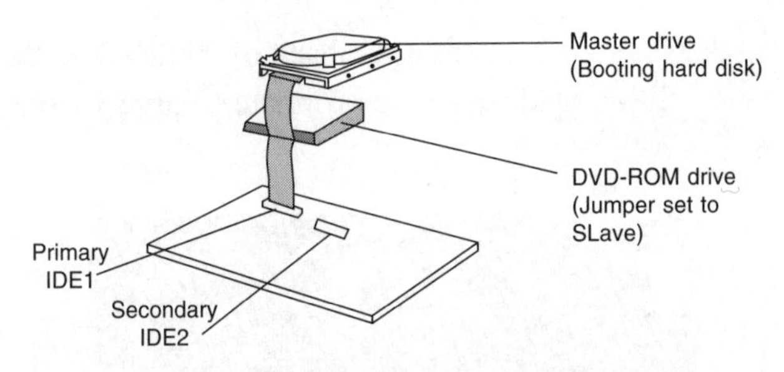

Figure 13.16a

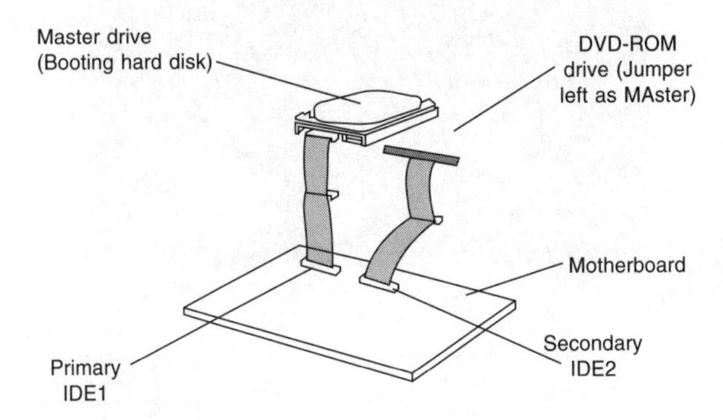

Figure 13.16b

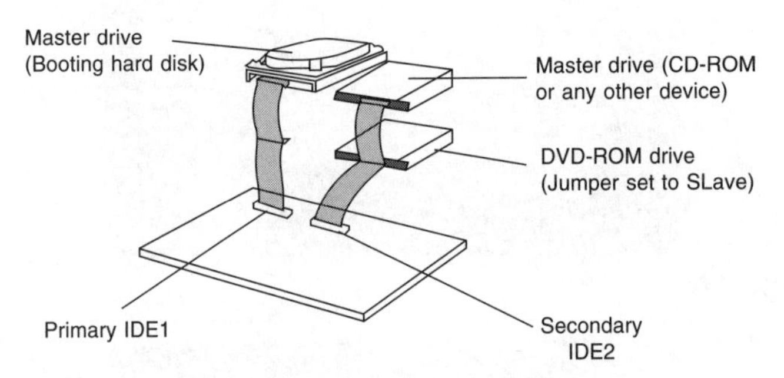

Figure 13.16c

As explained above, PCs have primary and secondary IDE connectors on the motherboard. The best place to hook up the DVD-ROM drive is on the channel that serves the CD-ROM drive.

If the secondary channel already has two drives (for example, both a CD-ROM drive and a CD-RW drive), connect the DVD-ROM drive to the free connector on the boot hard-disk drive's channel.

1. Make sure the jumper on the back of the DVD-ROM drive is set to Slave or Master as appropriate (Figure 13.17a)

Figure 13.17a

Figure 13.17b

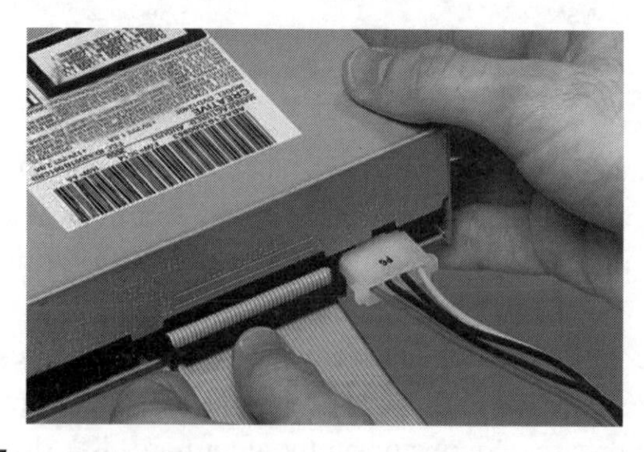

Figure 13.17c

2. If special brackets are needed to mount the drive in your PC, attach them now
3. Find a free mounting bay, remove the bay panel if necessary, and slide the DVD-ROM drive into it (Figure 13.17b)
4. Connect the ribbon data cable and the power cable to the rear of the DVD-ROM drive (Figure 13.17c)
5. Make sure the coloured edge (usually red) of the data cable is attached to pin 1 on the DVD drive connector (pin 1 is normally nearest to the power socket)
6. Connect the audio cable to the rear of the DVD-ROM drive and the other end to audio-in on the motherboard. Where a separate sound card is installed, insert the audio cable to the sound card.

Installing the drivers and DVD-ROM drive software

Hardware installation must be followed by installation of the relevant software. This involves installing the DVD-ROM driver and the drive application software driver.

Plug the PC back in and turn it on. Windows should detect the new drive, and will prompt you to instal a driver by launching the 'Add New Hardware Wizard' (Figure 13.18). If you are using Windows 98 or 2000, follow these steps to install the driver:

1. Click on the 'Next' button to begin the installation process
2. At the next prompt, select the 'Search for the best driver for your device' radio button and then click on 'Next'
3. The Wizard will prompt you for a location where the driver can be found; select 'CD-ROM'
4. Insert the CD-ROM provided by the DVD drive manufacturer into the CD-ROM drive, and then click on 'Next'
5. Windows will confirm that it has found the driver; click on 'Next'
6. The files will be copied and installed on the computer
7. Click 'Finish' to complete the installation.

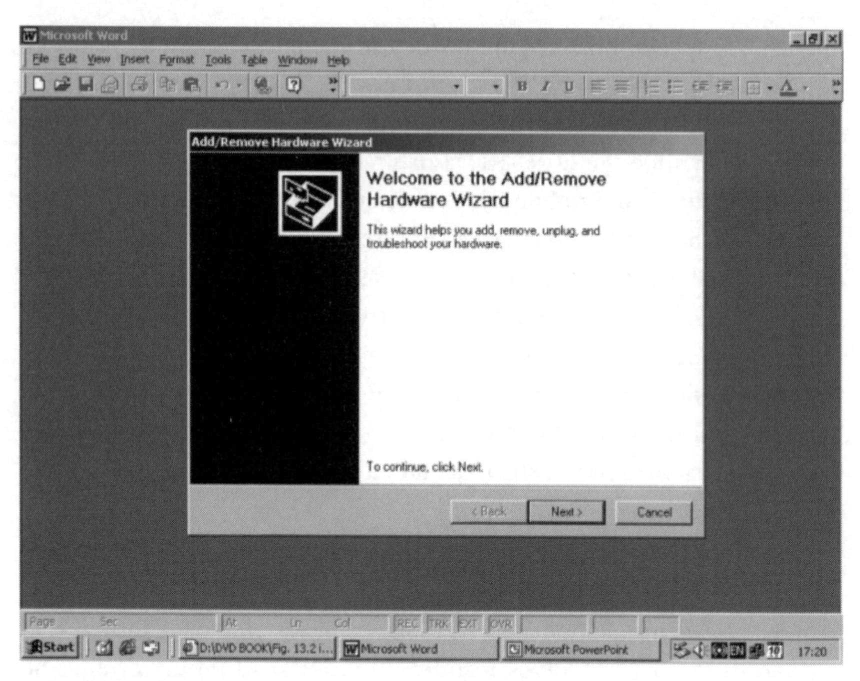

Figure 13.18

If your computer is running Windows ME or Windows XP, the following are the relevant steps to install the driver:

1. When the 'Add New Hardware' Wizard opens, select 'automatic search for a better driver'
2. Insert the CD-ROM disc supplied by the manufacturer into the CD-ROM drive, and click on 'Next'
3. Windows will search for the new driver and proceed to install it
4. When finished, Windows will inform you that the driver has been installed successfully
5. Click on 'Finish' to exit.

After the driver has been installed, the DVD-ROM drive software provided by the manufacturer must be installed. Windows 98 and 2000 will prompt you to install the drive software, and all you have to do is follow the on-screen instructions. If you need to install the software manually:

1. Select 'Start'
2. Choose 'Settings'
3. Select 'Control Panel'
4. Double-click on 'Add/Remove' programs icon, and follow on-screen directions.

Alternatively, the application software may be installed directly from the manufacturer's CD-ROM disc:

1. Insert the CD-ROM disc provided by the manufacturer in the drive, and click on 'Start'
2. Select 'Run'
3. Click on 'Browse'
4. Double-click the CD-ROM drive to open it
5. Select 'Setup.exe', and click on 'OK'.

The installation sequence will commence and the software will be installed on the hard disk in an appropriate folder.

To check that the device has been installed successfully, open the device manager in the following way:

1. Select 'Start'
2. Choose 'Settings'
3. Select 'Control Panel'
4. Double-click the 'System' icon
5. Select 'Hardware'
6. Double-click 'Device Manager'.

There should be no exclamation mark, question mark or cross against any device. If there is, then double-click on the device to ascertain the cause for the malfunction (e.g. resource conflict or driver not installed). To troubleshoot, click on 'Troubleshooter' and follow on-screen instructions.

Finally, test to be sure that the new drive will read both standard CD-ROM and DVD-ROM discs. If you have problems, turn off your PC and recheck all your connections.

Installing the MPEG decoder card

While MPEG decoding may be carried out by the processor using special video/audio decompression software, it is advisable to use dedicated hardware to carry out the complex and time-consuming processing operations involved in decoding video and audio DVD packets. The MPEG decoder card has a number of ports that are accessible from the back of the computer, and a typical set of ports is shown in Figure 13.19. These include:

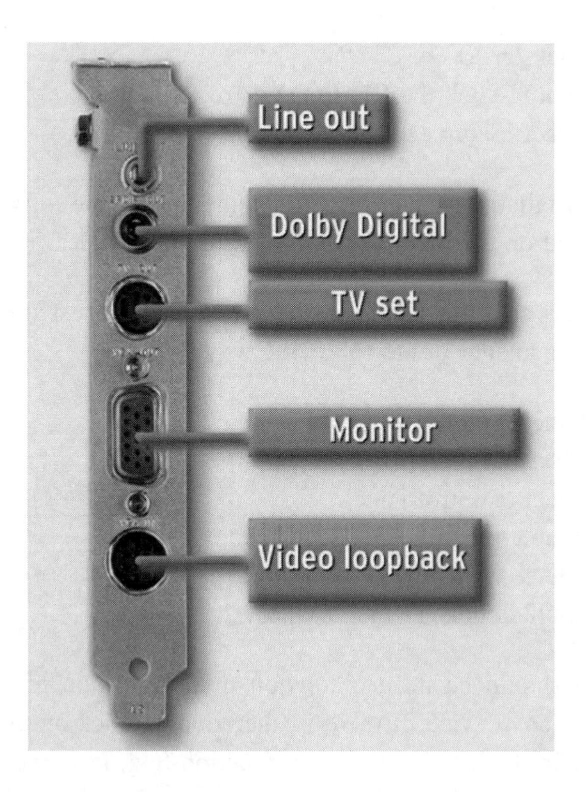

Figure 13.19

Line out	Stereo audio output for the sound card
Dolby Digital	Digital output to be fed to a Dolby Digital audio decoder for 5.1 surround sound
TV set	S-video output port for a TV receiver
Monitor	Video out for the monitor
Video loopback	Video input from the video card

Installing the MPEG decoder card involves the physical or hardware installation as well as software installation of the driver and relevant associated software. The step-by-step procedure for installing an MPEG card is listed below (refer to Figure 13.20):

1. Turn off the PC, unplug it, and remove the cover
2. Find a free PCI expansion slot and remove the metal cover on the back of the slot

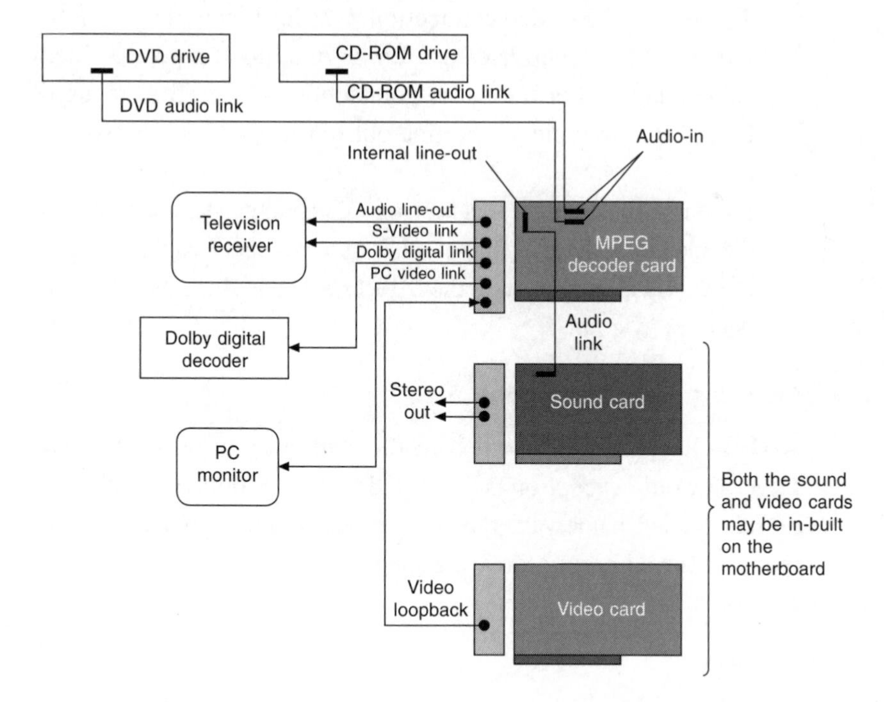

Figure 13.20

3. Carefully slide the card into the slot, making sure it's firmly seated, and secure it with a screw
4. Connect the 'audio out' from the DVD-ROM drive (and the CD-ROM drive if a separate drive is installed) to the internal 'audio-in' of the MPEG card
5. Connect an audio cable from the internal 'line out' connector of the MPEG card to the 'audio-in' on the motherboard. Where a separate sound card is installed, insert the audio cable to the sound card
6. Disconnect the monitor cable from the PC's video card and connect it to the monitor port on the MPEG card
7. Use a video loopback cable, normally included with the DVD-ROM drive, and connect it to the MPEG card and to the PC's video card
8. If you wish to watch DVD movies on a television set, you will need to make two connections: video and audio. Connect the TV receiver to the MPEG card using the S-video port if the TV set has an S-video connection. For this you will need an S-video cable. Otherwise, use an S-video-to-composite-video cable adapter. For the audio connection, you need a standard RCA cable to connect the line out to the TV receiver's audio input
9. If you have a stereo receiver or PC speaker system that can decode Dolby Digital Surround Sound, connect a cable between the Dolby Digital port on the MPEG card and the Dolby Digital decoder system.

Installing recording drives

A DVD-RW drive is installed in the same way as a DVD-ROM drive. The only exception is that an additional application software must be loaded, namely the recording program and any other software required for authoring, creating menus, etc.

DVD GLOSSARY

AC-3 Dolby's digital audio coding standard, also known as Dolby Digital, containing five separate tracks (three at the front and two at the back) and one additional subwoofer, a format known as 5.1.

Access time A measure of the speed of a storage device such as a memory chip. Defined as the time interval between the instant when an address appears to the address bus and data appearing on the data bus.

Active high Applies to hardware control signals that are active, i.e. perform their task when their logic state is HIGH (or logic 1).

Active low Applies to hardware control signals that are active, i.e. perform their task when their logic state is LOW (or logic 0). Normally, hardware control signals such as Reset, Enable and Chip Select are active low.

ADC Analogue-to-digital conversion, a process in which an analogue waveform is sampled at regular intervals. Each sample is then converted into a binary code that represents its amplitude.

Address A binary of hexadecimal number that defines the address of a memory location.

Address bus A number of parallel conductors whose simultaneous logic states indicate the address of a memory location.

AGC Automatic gain control used to vary the gain of the tuner or other amplifiers in order to keep the output constant.

Aliasing Distortion caused when an analogue signal is sampled at or below the Nyquist rate of 2× the highest analogue frequency.

ALU Arithmetic and logic unit. The part of the CPU that carries out the arithmetic and logic operations.

Anamorphic A term used to describe the representation of a wide-screen video image by squeezing it horizontally to fit into a conventional 4 : 3 aspect ratio. It may be stretched back by the decoder to fit into a widescreen 16 : 9 format.

Anchor picture Reference frame used when obtaining predicted frames, also known as I-frame.

ANSI American National Standard Institute.

APS Analogue Protection System, Macrovision's copy protection system for DVD-video that distorts the video output so that if recorded to VHS tape the playback is unwatchable.

Artefact An unintended, unwanted visual aberration in a video image. Artefacts can be caused by many factors, including data compression, readout or transmission errors, and film-to-video transformation.

ASCII American Standard Code for Information Interchange. Standard for representing text characters.

ASIC Application Specific Integrated Circuit (see also system-on-chip, SoC).

Aspect ratio The ratio between the width and the height of a TV screen. Traditionally an aspect ratio of 4 : 3 (1.33 : 1) is used. Widescreen television uses an aspect ration of 16 : 9 (1.78 : 1).

Audio_TS The directory where audio data are stored in a DVD title.

Authoring Part of the production process of a DVD title, which brings the individual elements of the title (video, audio, graphics, sub-pictures, etc.) into a unified whole with navigational pathways and program chains (PGC) linking the various parts.

Autoplay A feature allowing a DVD disc to play automatically when inserted.

Bandwidth The range of frequencies that a device can handle. Generally speaking, the wider the bandwidth of a device or a communication channel, the greater the transfer rate. In DVD applications, bandwidth refers to the maximum bit available rate that a system or unit can handle. For instance, the bandwidth of the video stream is given as 9.8.

BER Bit error rate. A measurement of the accuracy of a received bitstream stated in terms of the ratio of error bits to total number of bits.

B-frame Bi-directional frame. A predicted frame constructed by using both past and future predicted frames.

Bit A binary digit with values 0 or 1 used in binary numbers and codes.

Bit budgeting The process by which bit rate is allocated to the various elements (audio, video, subtitles, stills, etc.) of a DVD title.

Bitmap Representation of characters or graphics by individual pixels arranged in row (horizontal) and column (vertical) order.

Bit rate Measure of the speed of a data stream given in bits per second (bits/s, or bps). In DVD applications, the maximum available bit rate is known as the bandwidth.

Block An 8×8 matrix of pixels or DCT coefficients.

Block coding A coding technique where a block of data k is encoded into a longer codeword of n digits by adding redundancy bits n – k.

Bonding The process of joining two substrates to make a DVD disc.

Book A Specification for DVD physical format

Book B Specification for DVD-video format

Book C Specification for DVD-audio format

Book D Specification for DVD record once format.

Book E Specification for rewritable DVD format.

Browser The software that provides a user with the facility to view HTML pages on the Internet.

Buffer memory A storage area that provides an uninterrupted and constant-rate flow of data.

Burst A few cycles of the colour subcarrier incorporated within the composite video signal.

Burst cutting area An annular area within the DVD disc hub where a bar code can be written for additional information, such as serial numbers.

Burst errors Adjacent multiple error bits.

Byte A group of 8 bits.

Caddy Required to hold a disc before it is loaded into a player.

Card See Expansion card.

CAV Constant Angular Velocity. This refers to the rotation of the DVD disc. With CAV, the disc rotates at constant revolutions per minute. The alternative is constant linear velocity (CLV), in which the angular velocity of the changes to keep the speed of the track being read by the optical head constant. See also CLV.

CBR Constant Bit Rate. There are two methods by which video and other data may be encoded and decoded: CBR and variable bit rate

(VBR). With CBR, the bit rate allocated to the data (e.g. video) stream is constant throughout the encoding process. CBR also refers to the manner in which data is read from the DVD disc at a constant rate of 26.16 Mbits/s. See also VBR.

CCITT Comité Consultatif Internationale de Telegraphique et Telephonique, or Consultative Committee on International Telephone and Telegraphy, now known as the ITU-T (International Telecommunications Union – Telecommunication Standardization Sector). This is the primary international body for fostering co-operative standards for telecommunications equipment and systems.

CD Abbreviation for compact disc, an optical disc storage device.

CD-R CD-Recordable. Write-once recordable CD, a version of the compact disc that can be written to once only.

CD-ROM Compact Disc Read Only Memory. A version of the compact disc that allows computer data to be stored on a DVD disc.

CD-RW CD-Rewritable disc. A CD that can be written to and re-written a number of times.

Cell A unit of video (or audio) that can be as short as 0.5 s or as long as a complete film. Cells are the basic units for presentation of video (or audio) information, and can be grouped in different ways to share contents among different titles or skip or remove cells for parental lock or branching in the case of multiple angles.

Channel layer See user layer.

Chip set Several chips designed to work together to perform a defined operation.

Chrominance Colour components of a video signal.

CISC Complex Instruction Set Code. Processors that cannot carry out complex mathematical operations without a very large number of instructions, for example all Intel processors. See also RISC.

CLV Constant Linear Velocity. This is when the speed of the track being read by the optical head is kept constant by varying the angular velocity of the DVD disc. The advantage of CLV is that the data density remains constant through the recording track, optimizing the use of the surface area of the disc. See also CAV.

CNR Carrier-to-noise ratio. This provides a measure of the effect of noise on the received signal.

Coaxial digital output A connection for feeding pure digital audio along a coaxial cable.

Code rate The ratio of k : n where k is the number of digits of a code and n – k is the number of redundant bits added to the code for error correction capabilities.

Code-vector Another name for a codeword.

Codec Coder/decoder.

Codeword A valid code from among all available codes, also known a code-vector, or vector for short.

Colour difference This refers to the two signals used in colour television to represent the colour content of the video, namely R – Y and B – Y. Together with the luminance Y signal, the three components R, G and B of the original picture may be reproduced.

Combo drive A DVD-ROM drive which is capable of reading and writing CD-R and CD-RW as well as DVD–R and DVD–RW.

Composite video Analogue video signal that includes the video information together with blanking and sync pulses, also known as CVBS.

Compression The process of removing 'redundant' or unnecessary data from audio, video or any other digital information to produce a more compact form of data for storage on a disc or transmission purposes.

Copy protection A technique used on CD and DVD discs to prevent the contents being copied. Technologies used include watermarking, signatures on disc, and encryption.

CPU Central processing unit, normally referred to as the microprocessor.

CRC Cyclic Redundancy Check. A technique used to detect errors in a group of data bits.

CS Chip Select. A hardware chip control signal that enables the chip.

CVBS An analogue video signal consisting of composite video blanking and sync. Also known as composite video.

DAC Digital-to-analogue converter. A device that receives digital information in the form of a bitstream and converts it back into its original analogue form.

Data area The physical part of a DVD disc that contains the useful data stored on the disc.

Data bus A number of parallel conductors that together carry data information in and out of memory chips.

Data rate See bit rate.

dB Decibel, unit for measuring relative power.

dBm Decibel mW, unit of measuring power relative to 1 mW.

DCT Discrete Cosine Transform. The process of transforming the pixel values of a block into coefficients representing the spatial frequency

	components of the image represented by the block.
Decompression	Conversion of compressed data back into its original state.
Demodulation	The recovery of the signal information from the modulated carrier.
Demultiplexing	The separation of multiplexed signals. The reverse of multiplexing.
De-scrambling	A system for returning scrambled signals back to their original state.
DIMM	Dual in-line memory module.
Disc	Disc with a 'c' refers to an optical storage device such as a CD-ROM or a DVD disc.
Disc image	The complete set of files describing the logical and presentational data of a DVD title, formatted in DVD's UDF file system and ready to be created as a master.
Disk	Disk with a 'k' refers to a magnetic storage device such as a hard disk used in a computer system.
Display rate	The number of times a displayed picture is refreshed. Films have a refresh rate of 24 frames/s (fps), NTSC refresh rate is 30 fps, or 60 fps for interlaced displays. The corresponding figures for PAL are 25 fps and 50 fps. Computer displays may use higher display rates that may be set by software.
DMA	Direct memory access, a technique used in computerized systems for fast transfer of data in bulk without the interference of the processor.
Dolby Digital	See AC-3.
Dolby Pro-Logic	A multi-channel sound that is superior to Dolby Surround but not as good as Dolby Digital, offering five audio channels
Dolby Surround	Dolby's first multi-channel sound using four

audio channels: left, right, centre and surround. All channels are encoded onto two audio tracks.

Down-mix The conversion of multi-channel audio to a two-channel stereo. The DVD specification allows coefficients to be stored on disc to ensure that down-mixing results in the best quality stereo output.

DRAM Dynamic RAM. A type of a volatile read/write memory that requires regular refreshing of its contents. DRAM memory chips are used as the main memory store in DVD and other computerized systems.

DSP Digital Signal Processor. A programmable integrated circuit for carrying out fast processing of a range of tasks such as audio and video decoding.

DTS Digital Theatre System. The multi-channel audio coding used in cinemas and also on some DVD-video discs.

DTS Decoding Time Stamp. Part of the PES header that indicates the time at which a packet of video or audio should be decoded.

DTV Digital television.

DVB Digital Video Broadcasting. A European standard for broadcast, satellite and cable video transmission. It also used in DVD applications.

DVD Digital Versatile Disc, also known as digital video disc, a video/audio mass storage device using 8-cm or 12-cm optical discs.

DVD-5 Industry abbreviations for a DVD format (the number refers to the capacity in GB): single-sided, single-layer disc with a capacity of 4.7 GB (billions bytes).

DVD-9 Industry abbreviation for a DVD format (the

number refers to the capacity in GB): a single-sided, dual-layer disc with a capacity of 8.54 GB.

DVD-10 Industry abbreviation for a DVD format (the number refers to the capacity in GB): a double-sided, single-layer disc with a capacity of 9.4 GB.

DVD-14 Industry abbreviation for a DVD format (the number refers to the capacity in GB): double-sided with one single-layer and one dual-layer, giving a capacity of 13.24 GB.

DVD-18 Industry abbreviation for a DVD format (the number refers to the capacity in GB): double-sided, dual-layer disc with a capacity of 17.08 GB.

DVD-audio DVD-A, a pre-recorded DVD format intended to carry high quality audio data.

DVD–R, DVD+R Recordable DVD, a write-once DVD format, compatible with over 80 per cent of all DVD players and DVD-ROM drives, with a capacity of 2.6/4.7 GB per side.

DVD-RAM A rewritable format with the best recording features, used more as a removable storage device than a recording medium for audio or video information. DVD-RAM is not compatible with most DVD players and DVD-ROM drives. It currently has a capacity of 2.6/4.7 GB per side.

DVD-ROM The base disc format, which supports DVD-audio and DVD-video formats.

DVD–RW Rewritable DVD disc, the first DVD recording format released that is compatible with almost 70 per cent of all DVD players and DVD-ROM drives. It supports single-sided 4.7-GB and double-sided 9.4-GB disc capacities.

DVD+RW A rewritable DVD format with reputably better

features than DVD–RW, such as lossless linking and both constant angular velocity (CAV) and constant linear velocity (CLV) writing. It is compatible with almost all DVD players and DVD-ROM drives. It supports single-sided 4.7-GB and double-sided 9.4-GB disc capacities.

DVD-video DVD-V, a DVD format capable of carrying high quality video with multi-channel audio in up to three languages plus subtitles and menus to provide user interactivity. Other features include multiple camera angles, parental lock and random access.

ECC Error Correction Code: a code added at the end of a packet of data, which can be used to detect and correct errors that may have occurred during the process of transmission or, in the case of DVD applications, during the process of playback.

ECD Error correction and detection. See also ECC.

EEPROM Electrically Erasable PROM. EEPROM can be programmed and erased while in-circuit.

EFM Eight-to-Fourteen Modulation, which converts 8-bit words into 14-bit words. Used in audio CD applications.

EFM+ Eight-to-sixteen modulation, which converts 8-bit words into 16-bit words. Used in DVD applications.

EIDE Enhanced Integrated Drive Electronics. An improved IDE standard. See also IDE.

EN Enable. A hardware chip control signal similar to Chip select (CS), it makes the chip active.

Encryption A method of encoding information so that access by subscribers may be controlled.

Entropy coding A coding technique using variable-length lossless coding of a digital signal to remove redundant bits. Both MPEG-2 and Dolby Digital employ entropy coding. It is also used by digital theatre sound (DTS) and meridian lossless packing (MLP).

EOB End of block. A code used in DCT scanning to indicate that the remaining DCT coefficients of a block are all zero. Used in DCT encoding to reduce the number of bits required to describe a block of pixels.

EPROM Erasable PROM.

ES Elementary stream. Video, audio, sub-picture or other compressed data stream.

Expansion card Normally known as card, a printed circuit board that may be inserted in an expansion slot of a computer to perform a particular function, such as DVD decoder or modem.

FEC Forward error correction. The addition of redundant bits to a codeword so that errors introduced by the transmission or playback media may be detected and corrected.

Field In an interlaced video signal, the complete picture frame is divided into two fields, a top field containing odd scan lines and a bottom field containing the even scan lines.

File-system A defined way of storing files on a disc. A file system is necessary for all data storage media to allow data files to be accessed.

Flash A programmable memory store that retains its data when power is switched off.

Fourier transform A process for analysing a signal into its frequency components. Used to hold start-up routines in DVD and other computerized systems.

Frame A complete picture containing all scan lines. A frame contains two fields in an interlaced system.

Frame A single, complete picture in video or film recording.

GB Gigabyte. In computer applications, 1 GB = 1024 MB = 1024 × 1024 KB = 1 048 576 KB = 1024 × 1024 × 1024 bytes = 1 073 741 800 bytes. In other applications, 1 GB = 10^9 bytes = 1 billion bytes.

GOP Group of pictures, having the same reference picture frame (I-frame) and containing several P-frames or a combination of P- and B-frames. The number of frames in a GOP is determined by the DVD producer, and is typically 12.

GPIO General Purpose Input/Output.

Gray code A binary counting sequence that ensures that only one binary bit changes state when the count is incremented or decremented.

HD High definition. Used to describe pictures with more than 1000 scan lines.

HDTV High definition television employing more than 1000 scan lines.

Header The bytes in a data frame or packet that contain the identification and sync information.

Hex Hexadecimal. A numbering system with a base of 16. Hexadecimal numbers have 16 different characters: 0, 1, 2, 3, 4, 5, 6, 7, 8, 9, A, B, C, D, E and F. The alphabetical characters A, B, C, D, E and F represent 10, 11, 12, 13, 14 and 15 respectively.

HTML HyperText Mark-up Language. A language used for formatting text for transmission on the Internet.

Hz Hertz; cycles per second.

I/O Input/Output.

I^2C Interintegrated circuit. A two-line serial control bus used in microprocessor-based systems, with one line carrying the data and a second line carrying a clock pulse.

IDE Integrated Drive Electronics. A standard for connecting hard disk and other devices such as DVD drives to personal computers.

I-frame A reference frame for a group of pictures, GOP.

IM Inter-Metall. A three-line serial control bus used in microprocessor-based systems. One line carries the data, a second carries control enable signals, and the third carries the clock pulse.

Interactive video The combination of video and computer technology offering user interaction for training, games and other applications.

Interframe coding An MPEG compression technique that compares a frame with a previous frame (and sometimes with a following as well as a previous frame) to produce a motions vector representing the movement involved, and a difference frame known as the predicted frame. The predicted frame is then encoded and, together with the motion vector, can be used by a DVD player decoding system to recreate the original picture content.

Interlacing A video scanning technique that produces two fields for each complete picture frame: an odd or top field with odd scan lines, and an even or bottom field with even scan lines.

Interleaving A technique used to break up bursts of errors to improve error correction. There are two types of interleaving: bit interleaving, in which the order of the bits in a word are changed in a predetermined way; and block interleaving, in which the order of defined blocks of data is changed in a predetermined way.

Intra coding MPEG data compression within one frame, also known as spatial or DCT coding. It removes unnecessary repetitive information contained in a single frame.

Inverse telecine The reverse of 2 : 3 pulldown, where frames that have been duplicated to create 60 fps from 24 fps film are removed.

IRQ Interrupt request. A hardware interrupt used in microprocessor-based systems to request a service from the microprocessor. When a request is received, the microprocessor halts its operation and provides the service routine requested by the device. When that is finished, the processor goes back to its original program.

ISO International Standardization Organization. Worldwide group responsible for establishing and managing various standards, committees and expert groups, including several image-compression standards.

ISO 13818 ISO MPEG-2 standard.

JPEG ISO/CCITT Joint Photographic Expert Group, which defined a high-quality compression standard for still pictures using a DCT algorithm.

Land As opposed to a pit, a land is the raised area of the surface of a disc. Pits and lands represent the logic levels of the bits stored on the disc.

Laser Light Amplification by Stimulated Emission of Radiation. A means of generating coherent light that can be focused to a very small spot size ideal for reading compact discs, in laser beam recording, and for writing CD-R and DVD–R discs.

Layer 0 The lower (nearest to laser pickup head), semi-reflective layer of a dual-layer DVD disc.

Layer 1 The upper (furthest away from the laser pickup

head), fully reflective layer of a dual-layer DVD disc.

Lead-in The physical area (normally 1.2 mm) preceding the data area on a disc. It contains, among other things, control data and the Table of Contents.

Lead-out The physical area (normally 1.2 mm) that succeeds the data area. It is the last area of a disc.

Letterbox A method for displaying widescreen 16 : 9 video on a conventional 4 : 3 TV by adding black bars to the top and bottom so that the full width of the image is seen.

Linear PCM See LPCM.

Lossless compression A compression technique that can be decoded without any loss of information.

Lossy compression A compression technique that, when decoded, will result in a loss of some information.

LPCM Linear PCM (pulse code modulation), one of the audio coding formats that does not use data compression.

LPF Low Pass Filter. A device that allows low frequencies through while attenuating higher frequency signals.

LSB Least significant bit. The right-most bit of a binary number.

Luminance Black and white content of a picture.

Macroblock Picture space of 16 × 16 pixels.

Masking The exclusion of sound perception by the human ear due to the presence of other sounds.

Master clock A clock that is used to control all other clocks.

MB Megabyte. In computer-based applications, 1 MB = 1024 KB = 1024 × 1024 bytes = 1 048 576 bytes. In other applications, 1 MB = 10^6 bytes = one million bytes.

Middle area On a dual-layer disc, using the opposite track path (OTP) in which one layer is read in the opposite direction to the other, the middle area is the area adjacent to the outside of the data area.

MLP Meridian lossless packing, a lossless data compression technique used by DVD-audio that removes redundant data from pulse code modulated (PCM) audio. A compression ratio of 2 : 1 is achievable; however, a lower compression ratio of 1.6 : 1 is normally used. MLP is necessary because the bit rate required for a linear PCM audio is very high for most DVD applications.

Motion estimation The process of estimating motion vectors during the MPEG encoding process.

Motion vector A vector that represents the direction and speed of the estimated motion of a block of pixels.

MP@ML Main Profile at Main Level. An MPEG data compression technique for standard TV quality, using a resolution of 720 × 576. Used in DVD-video applications.

MP3 MPEG-1 layer III audio compression format used on the Internet for music files.

MPEG ISO/CCITT Moving Pictures Expert Group. Group that has defined MPEG-1, MPEG-2 and MPEG-4 video compression standards.

MPEG-1 ISO Moving Pictures Expert Group, designed for CD-ROM applications.

MPEG-2 ISO Moving Pictures Expert Group standard, designed for broadcast TV applications.

MSB Most significant bit. The left-most bit in a binary number.

Multi-angle The option supported by DVD to store (and for the user to select) scenes from up to nine different camera angles.

Multimedia Normally computer information in more than one form, such as video, audio, text, subtitles, still images and animation.

Multiple language The option supported by DVD to store and play up to eight different languages.

Multiple language subtitles The option supported by DVD to provide subtitles in up to 32 different languages.

Multiplexing The process of combining a number of independent signals to share a single transmission medium.

MUSICAM Masking Universal Sub-band Integrated Coding and Multiplexing. A technique employed in MPEG audio coding.

MUX Multiplexer.

NA Numeric aperture. A measure of the focusing power of a lens. A lens with a high NA has more focusing power than a lens with a lower NA value.

Navigation data Information that is used to provide search and branching facilities.

Noise Unwanted information added to a signal by the encoding process or the recording or transmission medium.

NRZ A technique for the representation of logical bits by a voltage level in which a pulse edge is introduced only when the data bits change from one state to another.

NTSC National Television System Committee. A television system used in the USA, having 525 lines and 30 pictures per second.

NRZI Non-return-to-zero invert, a technique for the representation of logical bits by a voltage level in which a pulse edge is introduced when data bits change and is inverted whenever there is a logic 1. Also see NRZ.

OE Output Enable. A hardware chip control

signal. It has the same effect as RD control signal.

Optical digital output A fibre optic connection for feeding pure digital audio signals from the DVD to external digital audio decoders.

OSD On-Screen Display. The displaying of various information on a TV screen, such as program number, date and other messages.

OTP Opposite Track Path. Where Layer 0 of a dual-layer DVD disc is read from the inner circumference to the outer circumference, and Layer 1 is read in the opposite direction – i.e. starting from the outside going inwards.

Overscan This refers to the scanning of a television picture in which the area scanned is slightly larger than the actual picture by about 4 to 5 per cent.

Pack A group of packets.

Packet A 2048-byte user data unit in a DVD stream. Packets are grouped in packs.

PAL Phase Alternate Line. A television system used in the UK and most other countries. With this system, the picture is scanned with 625 lines at 25 pictures per second.

Palette A table of colours from which a subset of colours may be identified. This technique allows fewer bits to be used to describe the colour of each pixel.

Pan & Scan A method for displaying widescreen video on a conventional TV by showing only part of the full width. The part shown is normally adjusted during the video, depending on where the most important action is.

Parental control See Parental lock.

Parental lock A means of preventing certain scenes on a DVD-video disc being seen by children.

Parity bit A method for error checking of a small string of data, normally 7 bits, using an extra bit known as the parity bit.

Payload Contents of a packet of data, other than the header.

PCI Peripheral Component Interconnect. A high-speed general purpose bus or slot provided on a motherboard of a personal computer for connecting peripheral devices such as DVD decoders, modems and sound cards.

PCM Pulse Code Modulation. A modulation technique that samples the analogue signal and then represents each sample by a binary code.

PCR Programme Clock Reference. A sample of the system clock at the transmitter is sent along the data stream in the form of a count, for the purpose of time synchronization at the receiving end.

Peripheral An attachment to a computer system, such as printer or keyboard.

PES Packetized Elementary Stream. An MPEG packet containing 2048-byte video, audio, sub-picture, DSI or PCI information.

P-frame Predicted frame produced as a result of comparing two successive frames.

PGC Program Chain. A group of cells or programs linked together to create a sequential presentation to the viewer. A program chain contains up to 99 programs.

PGCI Program Chain Information. Data contained in the PCI stream describing a chain of cells grouped in a program and their location.

PIC Presentation Information Control. A data stream providing information on the timing and presentation (aspect ratio, language and menu options) of a program.

PID Packet Identifier. A 13-bit code that is included in the header of the transport stream packet to identify the programme to which the particular packet belongs.

PIP Picture-In-Picture. A feature that shows a different video in a small window superimposed in the corner of the screen.

Pit A microscopic depression in the recording layer of a disc. Together with lands, pits represent data bits.

Pixel Abbreviation of picture element. The smallest element of a picture.

PLL Phase-Locked Loop. Used to ensure the synchronization of two clocks.

PnP Plug-n-play. A system developed by Intel so that operating systems can automatically recognize, instal and configure peripherals to a computer system.

Port address The address of a register to be used for the transfer of data between a computer and a peripheral.

PRBS Pseudo Random Binary Sequence. Used to remove uneven energy in digital streams.

Program A sequence of cells linked together to form a sequence of audio and video presentation.

Progressive scan A scanning method that scans all lines of a frame or picture sequentially – line 1, line 2, etc. – in contrast with interlaced scanning, which scans every other line.

PROM Programmable ROM. These are non-volatile memory chips that can have their contents deleted and a new program written by a special procedure.

PTP Parallel Track Path. Where Layer 0 of a dual-layer DVD disc is read from the inner circumference to the outer circumference and

Layer 1 is read in the same way – i.e. starting from the inside and going outwards.

PTS Presentation Time Stamp. This indicates the time a viewing unit must be presented to the viewer.

Pulldown A technique for converting 24-fps film to 30-fps NTSC video or 25-fps PAL by adding fields.

PWM Pulse Width Modulation. A modulation technique whereby the width of the pulse changes in accordance with the signal.

QEF Quasi-Error Free. A system is described as QEF if the bit error rate (BER) is less than one in 10^{11}.

Quantizing The process of determining the level of a sample within a given number of discrete levels.

Quantizing error An error in a digital system caused by the inherent ambiguity in the least significant bit.

RAM Random Access Memory. A volatile read/write memory store, which loses its data if power is switched off.

Raster The pattern of scan lines that makes up the television screen when no video is being fed into the receiver.

RD Read. A hardware chip control signal that instructs the chip to place data on its data bus.

Red Book The specification that describes the audio CD.

Redundancy The inclusion of extra bits to a code to improve error detection and correction at the receiving end.

Reed-Solomon An error detection and correction technique used in DVD applications.

Reference frame See I-frame.

Reference picture See I-frame.

Refresh rate	The number of frames displayed per second for film and video. For NTSC it is 30, and for PAL/SECAM it is 25.
Regional coding	Regarding DVD-video specification, the world is divided into six regions or locales so discs can be made to play in only one or a limited number of regions.
Resolution	Resolution. A measure of the total number of pixels available on a video display. Normally indicated as number of horizontal pixels × number of vertical pixels (e.g. 720 × 756).
REST	See RST.
RGB	Red–Green–Blue. The three primary colours used in video and television applications.
RISC	Reduced Instruction Set Code. Powerful and fast processors that are capable of carrying out complex mathematical operations with fewer instructions compared with CISC processors (for example, Oak processors). See also CISC.
RLC	Run-Length Coding. A coding method that encodes a series of identical bits as a group instead of individually.
Routine	A short program that is used frequently.
RST	Reset. A hardware chip control signal that resets the programmable chip, such as a CPU, and a video/audio decoder.
Sampling	The first step in the process of converting an analogue signal into a digital representation. This is accomplished by measuring the value of the analogue signal at regular intervals to give 'samples'. These values are then encoded to provide a digital representation of the analogue signal.
Sampling rate	Number of samples taken per second.
Scan rate	See Refresh rate.

SCART A 21-pin connector used for interconnection between TV receivers, VCRs, DVD players and other A/V systems. They provide RGB, composite video and stereo sound connections.

SCL Serial clock for I^2C serial control bus.

Scrambling A method of coding aimed at randomizing the bitstream to remove uneven energy content and to prevent interference from a repeating pattern from a neighbouring track. It is often confused or substituted for encryption, which is a technique for restricting access by subscribers.

SDA Serial data for I^2C serial control bus.

SDDS Sony Dynamic Digital Sound, introduced in August 1994 for the film industry, and designed to provide the best and highest quality sound presentation. SDDS is an optional audio coding format for DVD-video.

SDRAM Synchronous DRAM. DRAM memory store with improved access time (i.e. faster). Used as the main memory store in DVD and other computerized systems.

SECAM A colour television system, similar to PAL (256 line and 25 pictures per second), used in France and a few other countries.

Sector A group of data recorded on a disk. There are two types: a *recording* sector contains 4836 bytes or 38 688 channel bits, and a *user* sector (used in the decoding process) contains 2048 bytes plus error correction and header information.

Seek time The time it takes the optical head to move to the required place on the track.

SIP System-In-Package. A more highly integrated system-on-chip, resulting in fewer components and simpler designs. See also SoC.

Slot Also known as expansion slot. A physical bus connector provided by the motherboard of a computer for the use of expansion cards.

S/N ratio Signal-to-noise ratio. The ratio of wanted signal information to unwanted extraneous noise.

SoC System-on-chip. A chip that combines the core of a microprocessor with embedded memory I/O ports UART and external bus interface. System-on-chip devices carry out general processing tasks as well as dedicated processing operation such as transport demultiplexing.

Software A series of computer instructions in the form of routines or programs that can be loaded into system memory to perform a particular task.

Spatial compression Intraframe data reduction, which removes unnecessary repetition of video information within a single picture frame.

SRAM Static RAM. A fast type of RAM chip that does not require refreshing. Used in applications such as DVD decoders, where fast access to data is necessary.

Statistical multiplexing Multiplexing that allocates different time slots for different signals according to need.

Storyboard Part of the production process of a DVD title. It is a set of visual mock-ups that are used to define a title's contents and the sequence of events.

Stream Continuous flow of data bits.

Sub-carrier A carrier that falls within the spectrum of another carrier. In analogue television, colour information is used to modulate a sub-carrier of 4.41 MHz, which falls within the 5-MHz bandwidth of the monochrome video signal.

Sub-picture Overlay graphics image used for subtitles, menu highlighting and other purposes.

Surround sound A multi-channel audio system with speakers in front of and behind the listener, to create the feeling of being in an auditorium.

SVGA Super VGA, a video adaptor for a monitor capable of high resolutions, such as 600 × 1200, with 24-bit full or true colour of over 16 million colours.

S-video A video output composed of two separate signals; luminance and chrominance. Usually provided on yellow plastic connectors, S-video gives the highest video quality output.

Sync information Information to inform the video display unit of the start of a new scan line and the start of a new frame (for sequential scanning) or field (for interlaced scanning).

Telecine The process used to transfer film images to video, a process known as 2–3 pulldown for NTSC and 2–2 pulldown for PAL.

Temporal compression Interframe data reduction that removes similarities between successive frames. See also Interframe coding.

Tilt A measure of warping of an optical disc. Because of its high numeric aperture, tilt is of particular importance for DVD discs.

Title The largest presentation unit of a DVD disc, normally a complete movie.

Track The continuous spiral channel on a DVD disc. A track is also used to refer to a distinct portion of A/V information, such as a sound track for a specific language.

Track buffer In the DVD player, the track buffer is responsible for smoothing the fluctuations in the user bitstream coming from the pickup head caused by irregular disc accesses. It also performs the conversion between the constant transfer bit rate of the pickup head to the variable bit rate allocated to the various elements (namely video, audio, sub-picture, PCI and DSI packets).

Track pitch The distance between consecutive 'tracks' on a disc, measured in a radial direction.

Transfer rate The speed at which data is transferred to or read off a disc.

UDF Universal Disk Format. The file system used for DVD and MO disks.

ULSI Ultra Large Scale Integration. A chip with 100 000 active elements.

User data The data layer before the addition of formatting, interleaving and error correction information. In DVD applications user data are organized in sectors of 2048 bytes, each known as a packetized elementary stream (PES).

User layer Also known as the channel or physical layer, this is the data layer that is physically recorded on the disc in the form of 4836-byte sectors.

VBI Vertical Blanking Interval. The interval between the end of one field and the beginning of the next, during which the television screen is blanked out to allow the electron beam to return to the top of the screen. The interval covers lines 6–23 for PAL, and 10–23 for NTSC.

VBI packet Vertical Blanking Interval packet. This is directly inserted into the reconstructed video signal, and contains sub-pictures or captions. Compared with the sub-picture channel, which has 32 streams, there is only one VBI channel per program, with limited options in terms of colours and highlights.

VBR Variable Bit Rate. This refers to the encoding process of a DVD video bitstream, in which the allocation of the bit rate may be made to vary according to the contents of the picture. Fast-moving video frames are allocated a higher bit rate than slow-changing frames, keeping the bit rate low. Conversely, for a specified disc space, VBR invariably results in higher picture quality than CBR. VBR provides a more efficient use of the restricted bit rate in terms of improved picture quality. A similar process takes place in the decoding stage at playback, where the video stream is decoded at the same variable bit rate that it was encoded at.

VCO Voltage-Controlled Oscillator, in which the output frequency is determined by the DC voltage fed into it.

VES Virtual Enhanced Sound. An intelligent audio system that creates surround sound without rear speakers.

VGA Video Graphic Array. A video adaptor that supports 640 × 350 text, and graphics with up to 256 colours.

VHS Video Home System. The video cassette recorder that is now universally used for home video recording and playback, and is destined to be superseded by the far higher quality and performance of DVD-video.

Video manager The top-level menu linking multiple titles.

Video_TS The directory where DVD-video data are stored.

VLSI Very Large Scale Integration. A chip with up to 10 000 active elements.

VMGM Video Manager Menu.

VOB Video Object. Part of or a complete program

	stream containing MPEG video, audio and navigation data.
Volatile	A memory store that loses its data when its DC power supply is removed. Examples of such memory chips are RAM, DRAM and SDRAM.
VTS	A set of one to ten files holding the contents of a title.
VTS	Video Title Set.
White Book	Defines the video CD standard for up to 74 minutes of VHS-quality MPEG-1 video on one CD.
Widescreen	A video image that is wider than the traditional 4 : 3 aspect ratio. For DVD application, widescreen normally means an aspect ratio of 16 : 9.
WR	Write. A hardware chip control signal that instructs data to be written into the chip.
Y	Common abbreviation for the luminance or luma signal.
Yellow Book	Defines the CD-ROM specification.
YUV	A colour encoding scheme for pictures in which the luminance and chrominance components are separately encoded. The human eye is less sensitive to colour variations than to brightness variations. YUV allows the encoding of luminance (Y) information at full bandwidth, and chrominance (colour difference UV) information at reduced bandwidth.
Zigzag Scanning	Zigzag scanning. Sequence of scanning DCT coefficients so that those most likely to have zero value appear at the end of the scan.
Zoned CLV	Zone on DVD disc where a constant linear velocity is employed.
Zones	See Regional coding.

APPENDIX A: INTEGRATED CIRCUITS

Logic gates

A logic gate is a two-state device, i.e. it has a two-state output: an output of 0 volts representing logic 0 (or LOW), and a fixed voltage output representing logic 1 (or HIGH). The logic gate may have several inputs, all of which may be in one of the two possible logic states: 0 or 1. Logic gates may be used to perform several functions, e.g. AND, OR, NAND, or NOR.

The list of all possible combinations of the input and their respective outputs is known as the truth table of the gate. Figure A1.1 shows the British and international standard symbols, and Table A1.1 shows their truth table.

Table A1.1 Truth table

Inputs		*Output functions*					
A	*B*	*AND*	*NAND*	*OR*	*NOR*	*EX-OR*	*EX-NOR*
0	0	0	1	0	1	0	1
0	1	0	1	1	0	1	0
1	0	0	1	1	0	1	0
1	1	1	0	1	0	0	1

Logic packages

Logic elements, including gates and memory devices, are manufactured in integrated circuit (IC) packages. These ICs are classified into categories, known as families, according to the number of gates or equivalent elements that they contain. These families are:

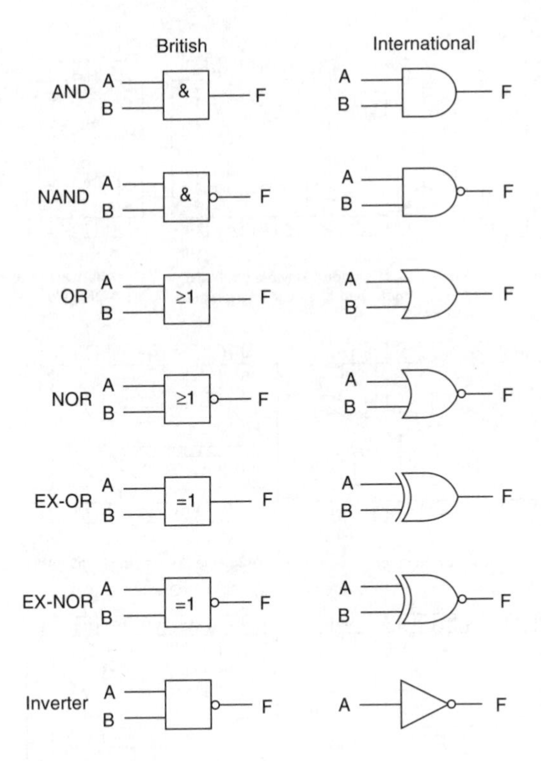

Figure A1.1
British and International gate symbols

Small scale integration (SSI)	Up to 10 gates
Medium scale integration (MSI)	10–100 gates
Large scale integration (LSI)	100–1000 gates
Very large scale integration (VLSI)	1000–10 000 gates
Super large scale integration (SLSI)	10 000–100 000 gates

The level of integration represents the complexity of the IC package, and increases in powers of 10, i.e. 10, 100, 1000, and so on. Small and medium scale integration (SSI and MSI) families provide discrete logic elements such as gates, counters and registers. Large and very large scale integration (LSI and VLSI) families provide memory chips, microprocessors, and complete systems such as 4-bit microcomputers.

Figure A1.3 shows a number of logic gate IC packages.

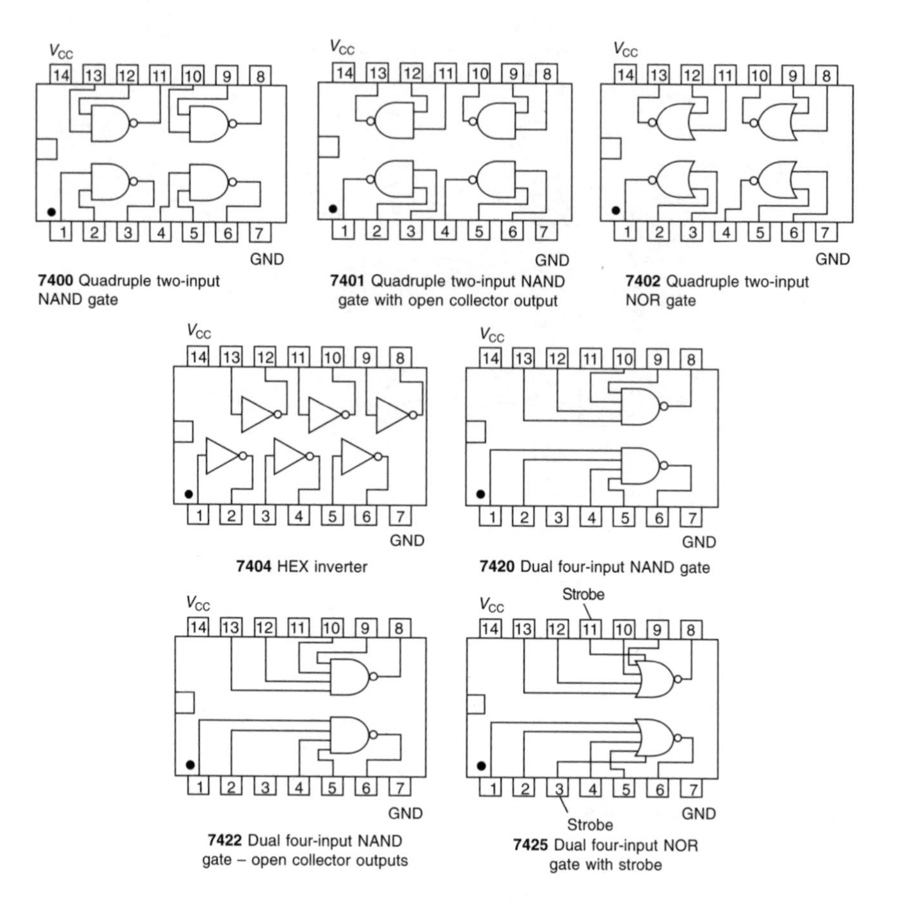

Figure A1.2

Other logic devices

Other devices that are commonly used in computerized systems are shown in Figure A1.4:

(a) Op amp
(b) Headphone amp
(c) DC regulator
(d) Line driver
(e) Flipflop and truth table
(f) DRAM block diagram, pinout and truth table
(g) SDRAM

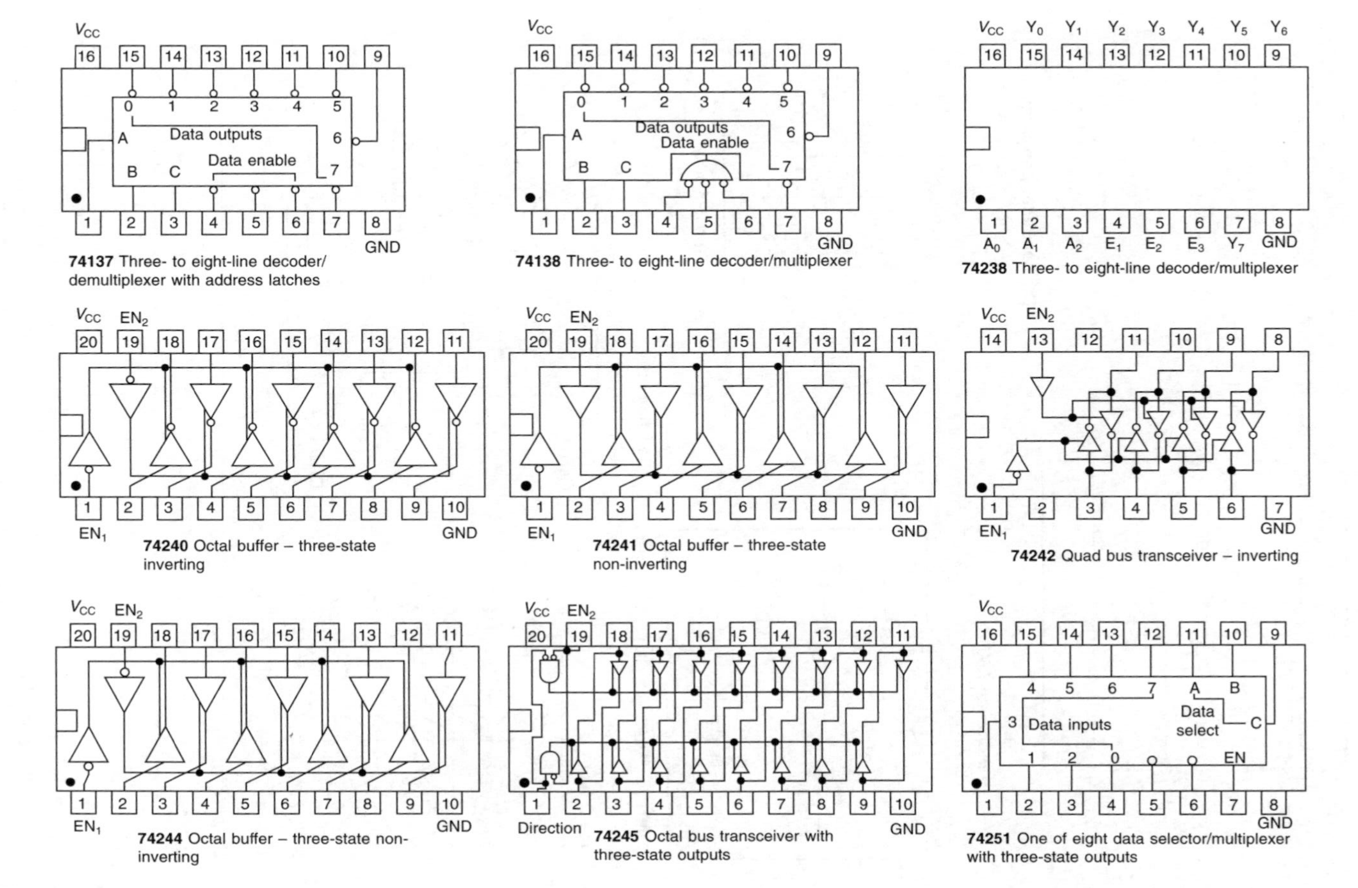
V_{CC}
Data outputs
Data enable
GND
74137 Three- to eight-line decoder/ demultiplexer with address latches
74138 Three- to eight-line decoder/multiplexer
Y_0 Y_1 Y_2 Y_3 Y_4 Y_5 Y_6
A_0 A_1 A_2 E_1 E_2 E_3 Y_7
74238 Three- to eight-line decoder/multiplexer
EN_2
EN_1
74240 Octal buffer – three-state inverting
74241 Octal buffer – three-state non-inverting
74242 Quad bus transceiver – inverting
74244 Octal buffer – three-state non-inverting
Direction
74245 Octal bus transceiver with three-state outputs
Data inputs
Data select
EN
74251 One of eight data selector/multiplexer with three-state outputs

Figure A1.3

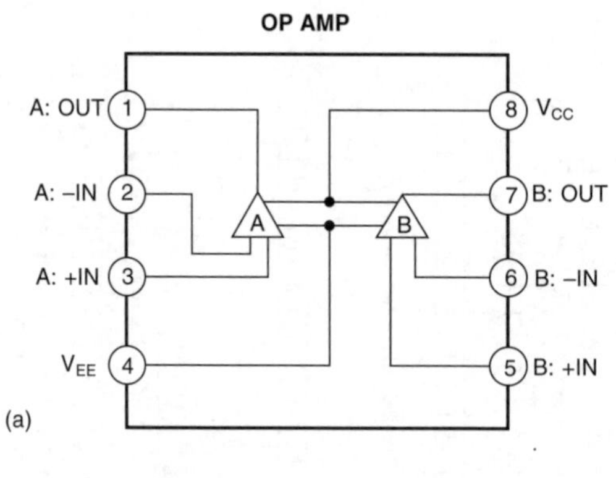

(a)

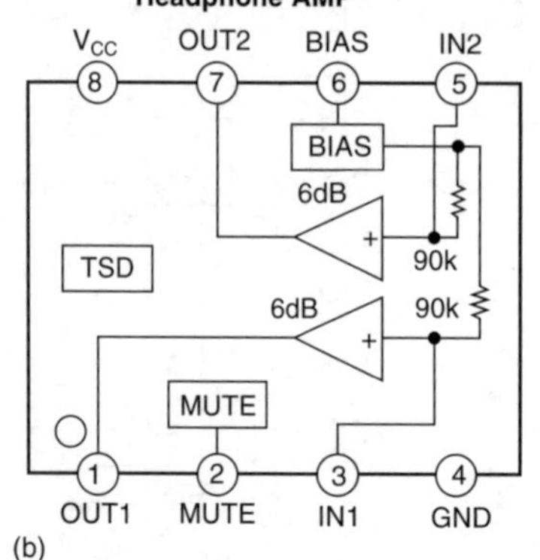

(b)

DC Regulator power supply

1. GND
2. Vc (on/off)
3. Vo
4. Vosense
5. Vin

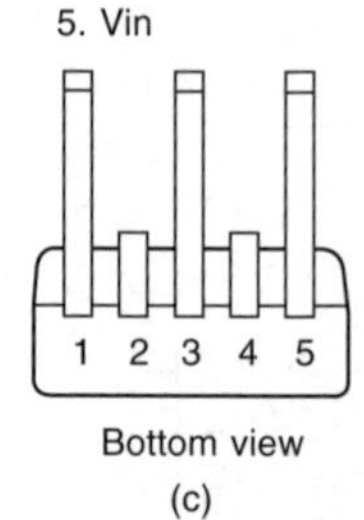

Bottom view

(c)

Line Driver

(d)

Flip - flop chip

Connection diagram

Truth table

Inputs				Output		Function
$\overline{CLR}$	$\overline{PR}$	D	CK	Q	$\overline{Q}$	
L	H	X	X	L	H	Clear
H	L	X	X	H	L	Preset
L	L	X	X	H(Note 1)	H(Note 1)	
H	H	L	↑	H	H	
H	H	H	↑	H	L	
H	H	X	↓	Qn	Qn	No change

(e)

Figure A1.4
Pin-out for specified chips

I²C controlled EEPROM chip

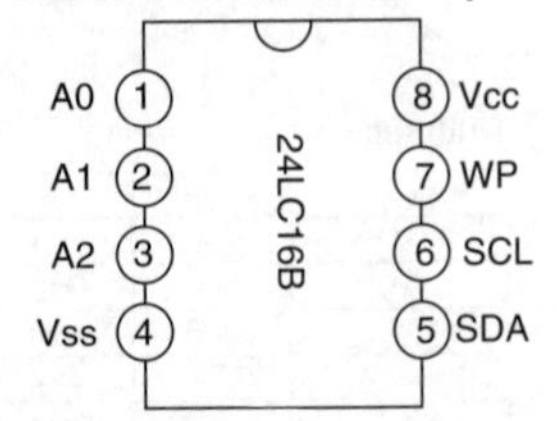

Truth Table

Name	Function
Vss	Ground
SDA	Serial Address/Data I/O
SCL	Serial clock
WP	Write Protect Input
Vcc	+2.5V to 5.5V Power Supply
A0.A1.A2	No Internal Connection

(f)

Dynamic RAM (DRAM)

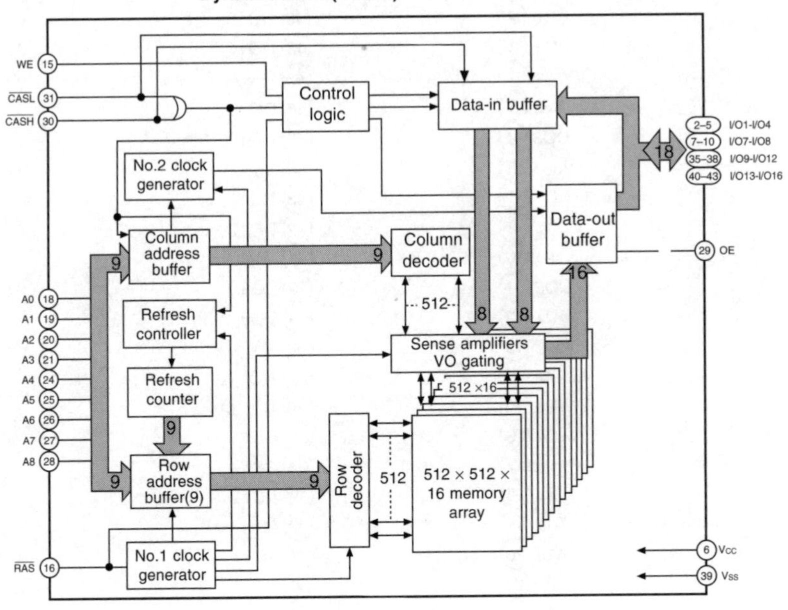

Block diagram

(g)

Figure A1.4
(Contd)

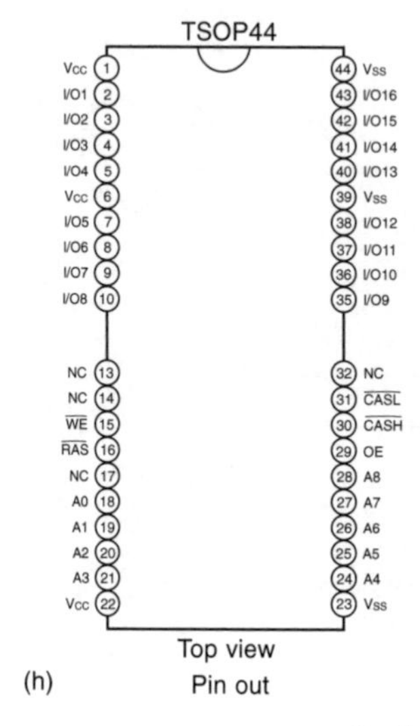

(h)

Truth table

Pin No.	SYM.	Type	Description
16-19,22-26	A0-A8	Input	Address Input
14	RAS	Input	Row Address Strobe
28	$\overline{CASH}$	Input	Column address strobe/upper byte control
29	$\overline{CASL}$	Input	Column address strobe/lower byte control
13	$\overline{WE}$	Input	Write Enable
27	$\overline{OE}$	Input	Output Enable
2-5,6-10,31-34,36-39	I/O1-I/O16	Input/Output	Data Input/Output
1,6,20	V_{CC}	Supply	Power, 5V
21,35,40	V_{SS}	Ground	Ground
11,12,15,30	NC	–	No connec

Synchronous DRAM (SDRAM) Pin out

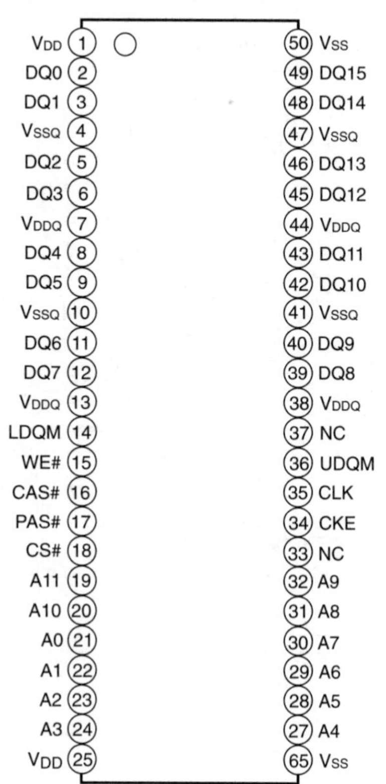

(i)

Figure A1.4
(Contd)

Appendix B: Required functions of a DVD-video player

There are a number of functions and operation methods that are required by the DVD-video format. As these functions are of fundamental importance to the system, they are designed into all DVD-video players. The various functions, and the reasons for them, are listed below.

Video playback

Composite video output

As the DVD disc can be recorded in the 16 : 9 aspect ratio, there is a requirement for this to be changed into the aspect ratio of a domestic television. Letterbox, 4 : 3 or panned options are selectable by the user.

Multi-angle playback function

With some DVD-videos, it is possible to change the camera angle at which the disc is played back. It is also possible to change the camera angle at certain points during playback, determined by the information laid down on the disc itself. There are two methods of angle change: seamless angle change and non-seamless angle change.

With non-seamless angle change, the video is paused at the switching point. The multi-angle method is specified for each angle block during production of the disc, and cannot be changed by the user.

Audio playback

Audio switching function

For discs with multiple audio tracks, the player allows these tracks to be switched during playback, depending on user operation. The DVD-video format allows common parts of video, audio, subtitles, etc. to be shared partially by several areas of the disc; this is called 'cell sharing'. For a shared cell, there is a feature to specify playable audio tracks of up to eight tracks for each title.

Linear PCM audio playback – two-channel analogue audio output

PCM sound is standard on all DVD-video discs. This can be a combination of 48/96-kHz sampling rates with 16-, 20- or 24-bit quantization. The player must be able to produce a minimum quality of sound at 48-kHz sampling with 16-bit quantization.

Other sound systems are optional, but for most Region 2 discs, the Dolby AC-3 system is used.

Subtitle playback

Subtitle selection

For discs with several subtitle tracks, the player allows the subtitle track to be switched during playback. In order to increase the recording efficiency of the disc, DVD-video allows common parts of video, audio and subtitle, etc. to be partially shared by several titles on the disc; this is called 'cell sharing'. For shared cells, there is a feature to specify playable subtitle tracks – up to 32 tracks for each title.

Display on/off function and forced display function

All DVD-video players have subtitle on/off functions, as well as a 'no sub/title selection' function. Regardless of the display on/off setting, the player will continuously decode the subtitle track during playback.

When the subtitle display is off, the player will decode the subtitle track for forced display functions. This will be in the same language as the currently selected subtitle track. When no subtitle track is selected, the player will decode it in the same language as the selected audio track. This means that when the disc software dictates that the display should be over-ridden or the on-screen menu displayed, it will be in the default language – i.e. the user will understand the messages displayed.

It is important that the preparation of the subtitle track(s) is correctly handled during the software production stage, to avoid any contradictions during playback.

Automatic selection playback function according to video aspect ratio

When subtitles corresponding to more than one aspect ratio are prepared for the disc, the player will automatically select the appropriate subtitle track that corresponds to that aspect ratio.

Playback control

Title selection and playback function

This is a function of the player to search and start playback of the selected title via the menu or direct key input.

Automatic set start function

This is a function of the player to start playback, according to the automatic set start procedure specified on the disc. With this type of disc, the user cannot select or start tracks other than those decided by the software producer. This is similar to the playback of a compact disc.

Menu call function

This is a function of the player to call the title selection menu, and the root menu recorded on the disc, under the control of the user.

The title selection menu can be called when the disc is stationary or just before the start of disc playback; the root menu can be selected after a certain title is selected.

Also called resume function, it is where the player resumes the original condition of the title playback if no effective user response is made when the menu is called during title playback. Depending on the existence of a subtitle while the title is being played back, the resumed playback position does not always match the position where the menu call is made.

Menu language select function

The player has a function to select a language when the systems menus are provided in several languages.

Additionally, it has a function to set the system menu default language. These selections are made while the player is in the stop condition. If there is no system menu in the system menu default language, the menu that is written in the default language of the disc will be displayed.

Menu operation function

All DVD-video players have selecting and activating functions for all menu operations as required.

Appendix C: DVD-video copyright protection systems

A new Hollywood movie released in the USA is usually released in Japan several months later, with the movie run schedule varying from one region to another in the world. Therefore, if the DVD-video software of a certain movie is put on sale in the USA before the end of the movie run in a different region, it is possible that imports of software may adversely affect the box-office earnings in that region.

To establish a system for tackling such potential problems, DVD-video divides the world into six sales regions and presets the DVD players and discs with the region ID numbers for areas where playback is permissible (Table C1.1 provides a list of all countries and their respective region IDs). A sales region ID number (e.g. 2 for Japan) is registered in each DVD player. In cases where a certain region ID number is recorded on a disc at the request of the copyright holder, the disc cannot be played back if the DVD player region ID number disagrees with the region ID number recorded on the disc. Of course there may be cases where the disc ID setup can be adjusted to permit playback in two or more regions. For the DVD software title of a past masterpiece, it is possible that the region ID setup may be adjusted to permit playback in all regions.

Table C1.1 Region ID numbers in different parts of the world

Afghanistan	5	Cocos (Keeling) Islands	4
Albania	2	Colombia	4
Algeria	5	Comoros	5
American Samoa	1	Congo	5
Andorra	2	Congo, the Democratic Republic of the	5
Angola	5	Cook Islands	4
Anguilla	4	Costa Rica	4
Antarctica	?	Cote D'Ivoire	5
Argentina	4	Croatia (Hrvatska)	2
Armenia	5	Cuba	4
Australia	4	Cyprus	2
Austria	2	Czech Republic	2
Azerbaijan	5	Denmark	2
Bahamas	4	Djibouti	5
Bahrain	2	Dominica	4
Bangladesh	5	Dominican Republic	4
Barbados	4	East Timor	3
Belarus	5	Ecuador	4
Belgium	2	Egypt	2
Belize	4	El Salvador	4
Benin	5	Equatorial Guinea	5
Bermuda	1	Eritrea	5
Bolivia	4	Estonia	5
Bosnia and Herzegovina	2	Ethiopia	5
Botswana	5	Falkland Islands (Malvinas)	4
Bouvet Island	?	Faroe Islands	2
Brazil	4	Fiji	4
British Indian Ocean Territory	5	Finland	2
Brunei Darussalam	3	France	2
Bulgaria	2	French Guiana	4
Burkina Faso	5	French Polynesia	4
Burundi	5	French Southern Territories	?
Cambodia	3	Gabon	5
Cameroon	5	Gambia	5
Canada	1	Georgia	?
Cape Verde	5	Germany	2
Cayman Islands	4	Ghana	5
Central African Republic	5	Gibraltar	2
Chad	5	Greece	2
Chile	4	Greenland	2
China	6	Grenada	4
Christmas Island	4		

Table C1.1 *(Contd)*

Guadeloupe	4	Macedonia, the Former Yugoslav Republic of	2
Guatemala	4	Madagascar	5
Guinea	5	Malawi	5
Guinea-Bissau	5	Malaysia	3
Guyana	4	Maldives	5
Haiti	4	Mali	5
Heard and McDonald Islands	4	Malta	2
Holy City (Vatican City State)	2	Marshall Islands	4
Honduras	4	Martinique	4
Hong Kong	3	Mauritania	5
Hungary	2	Mauritius	5
Iceland	2	Mayotte	5
India	5	Mexico	4
Indonesia	3	Micronesia, Federated States of	4
Iran, Islamic Republic of	2	Moldova, Republic of	5
Iraq	2	Monaco	2
Ireland	2	Mongolia	5
Israel	2	Montserrat	4
Italy	2	Morocco	5
Jamaica	4	Mozambique	5
Japan	2	Myanmar	3
Jordan	2	Namibia	5
Kazakhstan	5	Nauru	4
Kenya	5	Nepal	5
Kiribati	4	Netherlands	2
Korea, Democratic People's Republic of	5	Netherlands Antilles	4
Korea, Republic of	3	New Caledonia	4
Kuwait	2	New Zealand	4
Kyrgyzstan	5	Nicaragua	4
Lao People's Democratic Republic	3	Niger	5
Latvia	5	Nigeria	5
Lebanon	2	Niue	4
Lesotho	2	Norfolk Island	4
Liberia	5	Northern Mariana Islands	4
Libyan Arab Jamahiriya	5	Norway	2
Liechtenstein	2	Oman	2
Lithuania	5	Pakistan	5
Luxembourg	2	Palau	4
Macau	3	Panama	4
		Papua New Guinea	4
		Paraguay	4

Table C1.1 *(Contd)*

Peru	4	Svalbard and Jan Mayen Islands	2
Philippines	3	Swaziland	2
Pitcairn	4	Sweden	2
Poland	2	Switzerland	2
Portugal	2	Syrian Arab Republic	2
Puerto Rico	1	Taiwan	3
Qatar	2	Tajikstan	5
Reunion	5	Tanzania, United Republic of	5
Romania	2	Thailand	3
Russian Federation	5	Togo	5
Rwanda	5	Tokelau	4
Saint Kitts and Nevis	4	Tonga	4
Saint Lucia	4	Trinidad and Tobago	4
Saint Vincent and the Grenadines	4	Tunisia	5
Samoa	4	Turkey	2
San Marino	2	Turkmenistan	5
Sao Tome and Principe	5	Turks and Caicos Islands	4
Saudi Arabia	2	Tuvalu	4
Senegal	5	Uganda	5
Seychelles	5	Ukraine	5
Sierra Leone	5	United Arab Emirates	2
Singapore	3	United Kingdom	2
Slovakia (Slovak Republic)	2	United States of America	1
Slovenia	2	Uruguay	4
Solomon Islands	4	Uzbekistan	5
Somalia	5	Vanuatu	4
South Africa	2	Venezuela	4
South Georgia and the South Sandwich Islands	4	Vietnam	3
Spain	2	Virgin Islands (British)	4
Sri Lanka	5	Virgin Islands (US)	1
St Helena	5	Wallis and Futuna Islands	4
St Pierre and Miquelon	1	Western Sahara	5
Sudan	5	Yemen	2
Suriname	4	Yugoslavia	2
		Zambia	5
		Zimbabwe	5

Appendix D: Units and specifications

Prefixes and their values

Prefix	*Name*	*Common use (denary)*	*Computer use (binary)*
K or k	Kilo	$10^3 = 1000$	$2^{10} = 1024$
M	Mega	$10^6 = 1\ 000\ 000$	$2^{20} = 1\ 048\ 576$
G	Giga	$10^9 = 1\ 000\ 000\ 000$	$2^{30} = 1\ 073\ 741\ 824$
T	Tera	$10^{12} = 1\ 000\ 000\ 000\ 000$	$2^{40} = 1\ 099\ 511\ 628$

Relationship between disc capacity, bit rate and time

Disc capacity in GB

$$= \frac{\text{Disc capacity in Mbps} \times \text{time in hours} \times 60 \times 60}{\text{8 bits per byte} \times 1000}$$

Average bit rate in Mbits/s

$$= \frac{\text{Disc capacity in GB} \times \text{8 bits per byte} \times 1000}{\text{Time in hours} \times 60 \times 60}$$

Program time in hours

$$= \frac{\text{Disc capacity in GB} \times \text{8 bits per byte} \times 1000}{\text{Average bit rate in Mbps} \times 60 \times 60}$$

DVD formats

DVD-ROM	Read-only	*Book A*
DVD-Video	Video	*Book B*
DVD-Audio	Multi-channel	*Book C*
DVD-R	Recordable	*Book D*
DVD-RAM	Random access	*Book E*
DVD/RW	Rewritable	*Book F*

Quick reference

	Content	*Size*	*Maximum playback time*
DVD-video	Audio and motion picture	12 cm	Single-sided disc, approx. 4 hours Double-sided disc, approx. 8 hours
		8 cm	Single-sided disc, approx. 80 minutes Double-sided disc, approx. 160 minutes
Video CD	Audio and motion picture	12 cm	74 minutes
Audio CD	Audio	12 cm	74 minutes
		8 cm	20 minutes

Basic specifications of DVD

	Disc diameter (mm)	Disc thickness (mm)	Minimum pit length (μm)	Maximum pit length (μm)	Track pitch (μm)	Sector alignment	Reference scanning linear velocity (m/s)	File system	Modulation	Error correction	Readout wavelength of laser diode (nm) (reference)	NA of object lens (reference)	Data capacity (GB)
DVD-Video/ DVD-ROM (single-sided, single layer type)	120	1.2 (two 0.6-mm substrates bonded)	0.4	1.87	0.74	CLV	3.49	UDF Bridge (UDF & ISO 9660)	EFM plus (8–16)	RS-PC	650/635	0.6	4.7
DVD-video/ DVD-ROM (single-sided, dual-layer type)	120	1.2 (two 0.6-mm substrates bonded)	0.44	2.05	0.74	CLV	3.84	UDF Bridge (UDF & ISO 9660)	EFM plus (8–16)	RS-PC	650/635	0.6	8.5

Video and sound specifications for DVD-video compared with Video CD4 laserdisc

		DVD-video			Video CD	Laserdisc
Video	Video compression system	MPEG2 (MPCML)			MPEG1	
	Resolution (pixels)	720 × 576 pixels*			352 × 286 pixels*	Analog
	Horizontal resolution	Approx. 500 TV lines			Approx. 250 TV lines (same as VHS)	Approx. 420 TV lines
	Compression ratio	Approx. 1/40			Approx. 1/140	
	Video bit rate	9.8 Mbits/s, max. (variable)			1.15 Mbits/s (fixed)	Analog
	Field/frame	Field/frame			Frame	
	Aspect ratio	4 : 3/16 : 9 (pan & scan/letter box)			4 : 3	4 : 3
Audio	Audio	Eight streams, max.†			2 channels (stereo)	Analog 2 channels, digital 2 channels (16-bit/44.1 kHz) or analog 1 channel, Dolby Digital (AC-3: 1 stream, digital 2 channels (16-bit/44.1 kHz))
	Audio system	MPEG	Dolby Digital (AC-3)	Linear PCM	MPEG1 layer II	
	Audio bit rate	Max 912 kbits/s (per stream)	Max 448 kbits/s (per stream)	Max. 6.144 Mbits/s (per stream)	224 kbits/s (fixed)	
	Number of channels	Max. 7.1 channels/stream	Max. 5.1 channels/ stream	Max. 8 channels/ stream	2 channels only	
	Quantization bit sampling frequency	48 kHz	48 kHz	16-bit, 20-bit, 24-bit, 48 kHz, 96 kHz	16-bit, 44.1 kHz	
Others	Subtitles	2 bits, run length bitmap system, 32 streams, max.			Open caption only	Open caption, closed caption

*In the case of PAL, DVD-video is not compatible with the high definition system.
†Dolby Digital, MPEG or Linear PCM can be selected for each audio system.

Comparison between DVD and CD specifications

	DVD	*CD*
Wavelength	650 ± 5 nm	780 ± 10 nm
NA	0.60 ± 0.01	0.45 ± 0.01
Polarization	Circular	–
Light intensity at the rim (of the pupil of the objective lens)	RAD: 60–70%; TAN: ≥ 90%	≤ 50%
Surface aberration	≤ 0.033 λ RMS	≤ 0.07 λ
Laser diode noise	≤ –134 dB/Hz	–
Measuring scanning velocity	3.49 ± 0.03 m/s (single layer) 3.84 ± 0.03 m/s (dual layer)	–
Disc clamping force	2.0 ± 0.5 N	1–2 N
Circuit characteristics	Standard servo, equalizer, PLL, CLV servo, slicer characteristics	–
Track pitch	0.74 ± 0.01 μm (average) 0.74 ± 0.03 μm (instantaneous)	1.6 ± 0.1 μm
Eccentricity (radial runout)	100 μm peak-to-peak	140 μm peak-to-peak
Warpage (surface deviation)	± 0.3 mm	± 0.5 mm

DVD-RAM specifications

Recording media	Phase change disc
Disc dimensions	20 mm diameter, bonded 0.6-mm thick substrates
Laser wavelength, NA	650 nm, 0.6
Modulation method	8/16 encoding
Sector size	2 kB
ECC block size	32 kB
Error correction method	RSPC
Capacity	2.6 GB (one-sided)
Disc format	ZCLV (Zoned Constant Linear Velocity)
Number of data area zones	24 zones
Addressing method	CAPA (Complimentary Allocated Pit Addressing)
Recording bit length	0.41 μm/bit
Recording method	Mark edge recording
Track pitch	0.74 m, land and groove
Tracking method	Push-pull
File system	,t,c,e
User data rate	11.08 Mbits/s

Comparison of basic playback specifications with those of DVD-ROM discs

DVD standard	*Single-layer DVD/ROM*	*DVD–R*	*DVD–RW*	*Dual-layer DVD-ROM*
Laser wavelength	635/650 nm			
Objective lends NA	0.60			
Reflectivity	45–85%			18–30%
Modulated amplitude	0.60 min			
Data track form	Single spiral track			
Track pitch	0.74 μm			
Tracking method	DPD (Differential Phase Detection)			
Minimum pit length	0.40 μm			0.44 μm
Data modulation	8/16, RLL (2,10)			
Error correction	RSPC (Reed-Solomon Product Code)			
Channel bit rate	26.16 Mbits/s			
Scanning velocity	3.49 m/s (CLV)			3.84 m/s (CLV)
User data capacity	4.70 GB/side			4.25 GB/layer

DVD disc storage capacity

Name	*Type*	*Capacity in billions of bytes*	*Capacity in binary gigabytes*	*Approximate recording time in hours of video*
12-cm discs				
DVD-5	SS/SL	4.7	4.38	2.25
DVD-9	SS/DL	8.54	7.98	4
DVD-10	DS/SL	9.4	8.75	4.5
DVD-14	DS/DL on one side only	13.24	12.33	6
DVD-18	DS/DL on both sides	17.08	15.9	8

8-cm discs			
SS/SL	1.43	1.36	0.75
SS/DL	2.6	2.48	1.25
DS/SL	2.85	2.72	1.5
DS/DL	5.19	4.95	2.5

SCART socket pins and their functions

Pin	*Function specification*	*Signal*
1	Right channel audio out	0.5 V into 1 kΩ
2	Right channel audio in	0.5 V into 10 kΩ
3	Left channel audio out	0.5 V into 1 kΩ
4	Audio earth	
5	Blue earth	
6	Left channel audio in	0.5 V into 10 kΩ
7	Blue in	0.7 V into 75 Ω
8	Source switching (9–12 V)	Not specific, but usually max. 12 V into 10 kΩ
9	Green earth	
10	Intercommunication line	
11	Green in	0.7 V into 75 Ω
12	Intercommunication line	
13	Red earth	
14	Intercommunication line earth	
15	Red in	0.7 V into 75 Ω
16	Fast RGB blanking	varies (1–3 V)
17	CVBS earth	
18	Fast blanking earth	
19	CVBS out	1 V into 75 Ω
20	CVBS in	1 V into 75 Ω
21	Socket earth	

Summary of bit-rate breakdown of a DVD player

Layer	*Bit rate (million bits/s)*
Channel rate (8/16 modulation + sync words)	26.16
After 8/16 demodulation	13.00
Error correction bytes	2.00
User data bit rate	11.08
Information file + DSI packet	1.00
Mux_rate	10.08
Sum of video, audio and subpicture	9.80 (max.)

Bit rate for various DVD elements

DVD element	*Bit rate (Mbits/s)*	*Size (Mbits/s)*
Audio track Dolby Digital 5.1	0.384 or 0.448	
Audio track Dolby Digital stereo	0.192	
Audio track MPEG-1	0.192–0.384	
Audio track MPEG-2	0.384–0.912	
Audio track linear PCM (no data reduction)	6.114	
Audio track linear PCM (with MLP data reduction)	3.668	
Sub-picture per track	0.01	
Subtitle	0.04	
Still image		1.00
MPEG-2 video	Up to 9.8	

Video stream specification

Data compression method	*MPEG2, MPEG1*
Bit rate	9.8 Mbits/s max. (MPEG-2)
1.856 Mbits/s max. (MPEG-1)	
GOP size	36 fields max.
Screen display	
TV systems	525/60, 626/50
Aspect ratios	4 : 3, 16 : 9
Modes	Pan & scan, letterbox
User data	Closed caption

Audio stream specification

	Linear PCM	*Dolby Digital™*	*MPEG audio*
Fs	48 kHz, 96 kHz	48 kHz	48 kHz
Qb	16/20/24 bits	Compressed	Compressed
Bit-rate (per 1 stream)	Max. 6.144 Mbits/s	Max. 448 kbits/s	Max. 912 kbits/s

Sub-picture stream specification

Data format	*Run-length encoding, two bits per pixel*
Data size per picture	52 kB max. 720 × 480 (525/60)
Bit rate	Up to 3.36 Mbits/s
Colours	4 of 16

Linear PCM audio specifications

Sampling frequency	*Quantization bit*	*Max. number of channels (and bit rate)**
48 kHz	16-bit	8 (6.144 Mbits/s)
	20-bit	6 (5.760 Mbits/s)
	24-bit	5 (5.760 Mbits/s)
96 kHz	16-bit	4 (6.144 Mbits/s)
	20-bit	3 (5.760 Mbits/s)
	24-bit	2 (4.608 Mbits/s)

*Per stream; maximum bit rate per stream is 6.75 Mbits/s.

Available disc space for different-sized DVD discs (allowing for 4% for overheads)

Disc type	*Total capacity*		*Overheads*		*Available capacity*	
	GB	*Mbits*	*GB*	*Mbits*	*GB*	*Mbits*
DVD-5	4.7	37 600	1.88	1504	4.512	36 096
DVD-9	8.54	68 320	3.416	2733	8.1984	65 587
DVD-10	9.4	75 200	3.76	3008	9.024	72 192
DVD-14	13.24	105 920	5.296	4237	12.71	101 683
DVD-18	17.08	136 640	6.832	5466	16.397	131 174

DVD drive speed and data rate

Drive speed	*Data rate*
1×	11.08 Mbits/s (1.32 MB/s)
2×	22.16 Mbits/s (2.64 MB/s)
4×	44.32 Mbits/s (5.28 MB/s)
5×	55.40 Mbits/s (6.60 MB/s)
6×	66.48 Mbits/s (7.93 MB/s)
8×	88.64 Mbits/s (10.57 MB/s)
10×	110.80 Mbits/s (13.21 MB/s)
16×	177.28 Mbits/s (21.13 MB/s)
20×	221.60 Mbits/s (26.42 MB/s)
24×	265.92 Mbits/s (31.70 MB/s)
32×	354.56 Mbits/s (42.27 MB/s)

DBV profiles/levels

Profile/level	*Simple*	*Main*	*SNR scalable**	*Spatial scalable*	*High*
High/1920 × 1080 × 30 or 1920 × 1152 × 25†		MP@HL US digital HDTV			HP@HL
High-1440/ 1440 × 1080 × 30 or 1440 × 1152 × 25†		MP@ H1440		SSP@ H1440 European digital HDTV	HP@H1440
Main/720 × 480 × 29.97 or 720 × 576 × 25†	SP@ML digital transmission cable TV	MP@ML‡ DVD-video, Digital satellite broadcasting (PerfecTV and others)	SNP@MP		HP@ML
Low/352 × 288 × 29.97†		MP@LL	SNP@LL		

MPEG1 packet header structure

Field	*Definition*	*No. of bits*
Packet start code	000001BA (hex)	32
Bits 0010	Beginning of SCR field	4
SCR	System clock ref. (4 MSB)	3
Marker bit	Always set to 1	1
SCR	System clock ref. (15 intermediate bits)	15
Marker bit	Always set to 1	1
SCR	System clock ref. (15 LSB)	15
Marker bit	Always set to 1	1
Marker bit	Always set to 1	1
Mux rate	MPEG multiplex bit rate (multiples of 50 bytes)	22
Marker bit	Always set to 1	1

MPEG2 packet header structure

Field	*Definition*	*No. of bits*
Start code	000001 (hex)	24
Stream ID	PES identification	8
Packet length	Length of packet (bytes) after start code and stream ID	16
PES scramble control	Indicates whether the PES is scrambled	2
Flags	Various flags	14
PES header length	Length of the remaining part of the PES (x + y)	8
PES header sub-fields	Variable field depending on flags	x bytes
Stuffing	Optional	y bytes

Transport packet header

Field	*Definition*	*No. of bits*
Sync byte	Synchronization byte 10000111 (47 hex)	8
El	Transport error indicator (indicates errors from previous stages)	1
PUSI	Payload unit start indicator (start of PES in packet)	1
TRR	Transport priority	1
PID	Packet identifier	13
SCR flags	Transport scrambling flag	2
AF	Adaptation field flat	1
PF	Payload present flag	1
CC	Continuity counter (truncated PES portions)	4

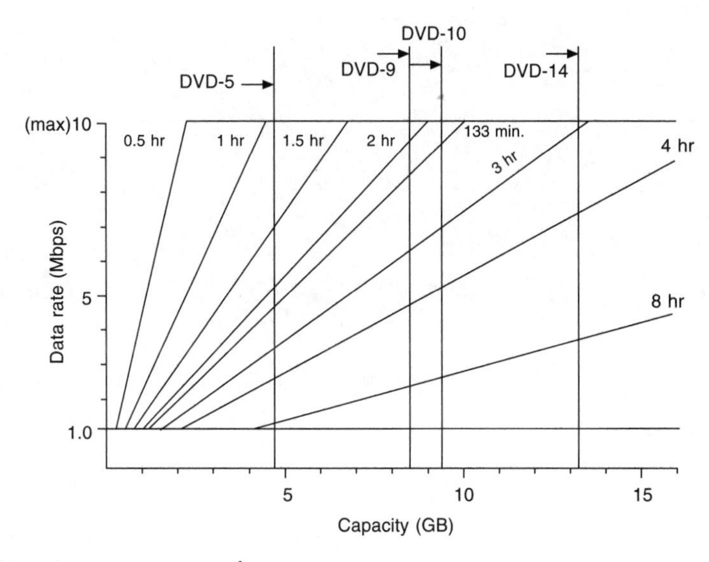

Data rate versus capacity

Allocation of interrupt requests in a computer

IRQ0	System timer
IRQ1	Keyboard controller
IRQ2	Cascade
IRQ3	Serial port COM2/COM4
IRQ4	(Shared) serial ports COM1/COM3
IRQ5	Available
IRQ6	Floppy disk controller
IRQ7	(Shared) parallel ports LPT1/LPT3
IRQ8	Real-time clock
IRQ9	Video unit
IRQ10	Available
IRQ11	Available
IRQ12	Available
IRQ13	Maths processor
IRQ14	Hard disk drive controller IDE1
IRQ15	Available

APPENDIX E: SELF-TEST QUESTIONS AND ANSWERS

Questions

1. Give a brief explanation of the following:
   a. Anamorphic
   b. AC3
   c. GOP
   d. PCI
   e. System-on-chip (SoC).

2. With reference to MPEG-2
   a. Briefly explain (i) temporal data compression; (ii) spatial data compression
   b. State the sampling frequency for (i) luminance; (ii) chrominance.

3. With reference to the pickup head, briefly explain what is meant by:
   a. Constant Linear Velocity (CLV)
   b. Constant bit rate (CBR)
   c. Focus search
   d. Photodetector.

4. Regarding a DVD:
   a. Calculate the required capacity of a DVD to store a 2-hour film, given an average transfer rate of 3.5 Mbits/s
   b. Explain briefly why the capacity of a double-layer DVD disc is *not* twice that of a single-layer DVD disc.

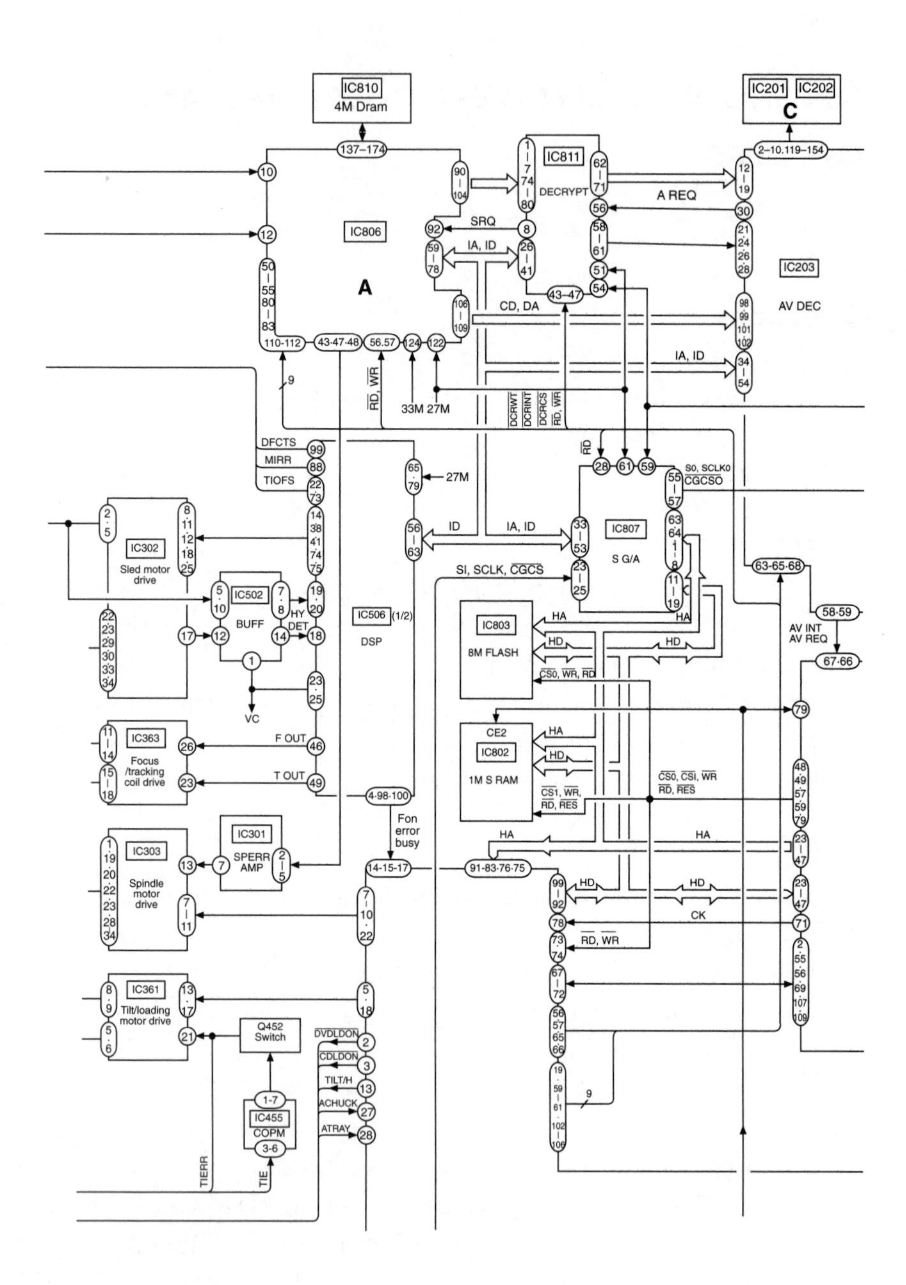
IC810
4M Dram
IC201
IC202
C
IC811
DECRYPT
A REQ
SRQ
IA, ID
IC806
A
CD, DA
IC203
AV DEC
IA, ID
33M 27M
DFCTS
MIRR
TIOFS
27M
IC807
S G/A
IC302
Sled motor drive
IC502
BUFF
HY DET
IC506 (1/2)
DSP
SI, SCLK, CGCS
IC803
8M FLASH
AV INT
AV REQ
VC
IC363
Focus /tracking coil drive
F OUT
T OUT
CE2
IC802
1M S RAM
Fon error busy
IC301
SPERR AMP
IC303
Spindle motor drive
CK
RD, WR
IC361
Tilt/loading motor drive
Q452
Switch
DVDLDON
CDLDON
TILT/H
ACHUCK
ATRAY
IC455
COPM
TIERR
TIE

Figure E1.1

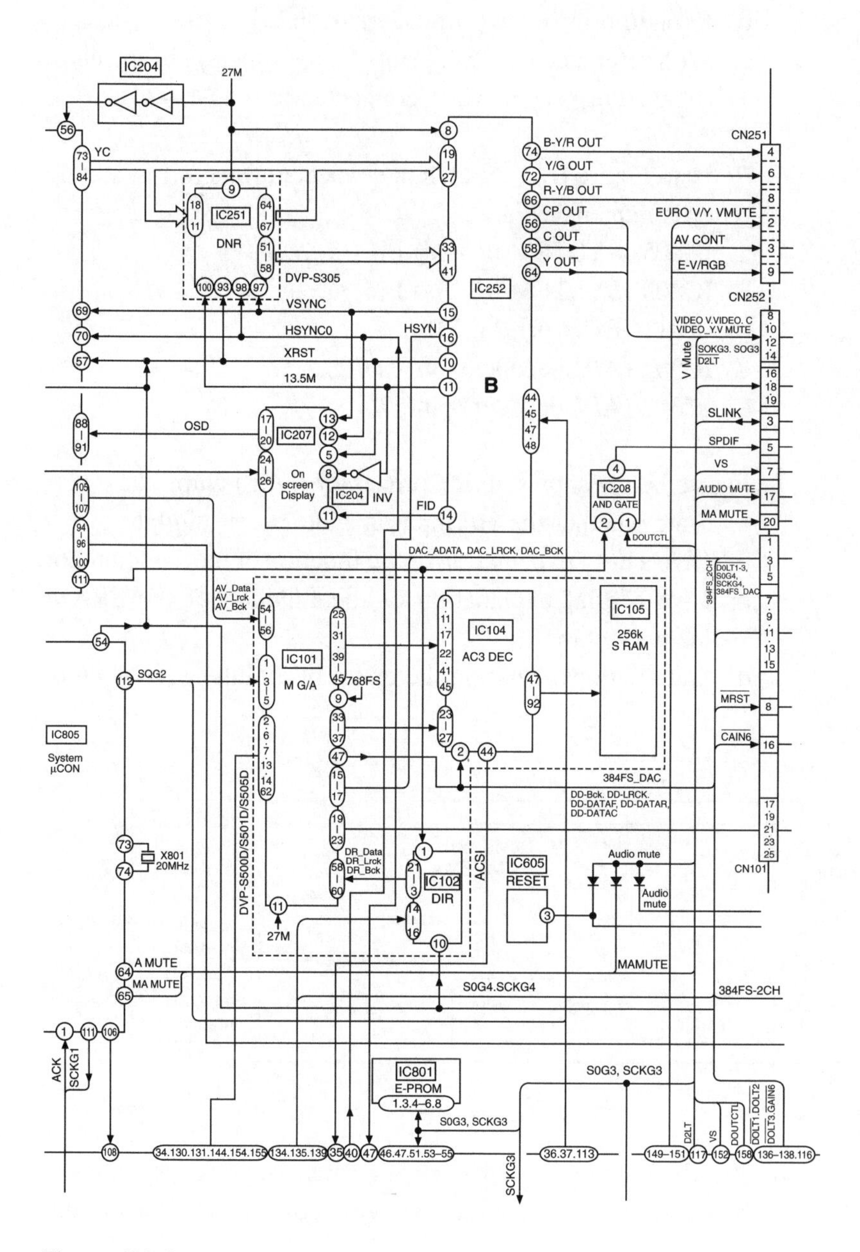

Figure E1.1
(Contd)

5. Referring to Figure E1.1:
   a. State the function of (i) Block A; (ii) Block B; (iii) Block C
   b. Explain briefly the purpose of IC803 (Flash)
   c. With reference to IC805, state (i) the function of the chip; (ii) the purpose of X801 connected to pins 73–74.

6. Refer to Figure E1.1. Sketch the expected waveforms, stating typical frequencies, at:
   a. IC805 (SYSTEM µCON) Pin 73
   b. IC805 (SYSTEM µCON) Pin 54
   c. IC811 (Decrypt) Pin 51
   d. IC302 (A/V decoder) pin 70
   e. IC302 (A/V decoder) pin 57.

7. Figure below shows an RF processor (DSP) chip.
   a. State whether the HF input is analogue or digital
   b. Give a brief explanation of the function of the demodulator
   c. Give a brief explanation of the function of the SRAM memory chip
   d. State with explanation the effect of a faulty SRAM chip.

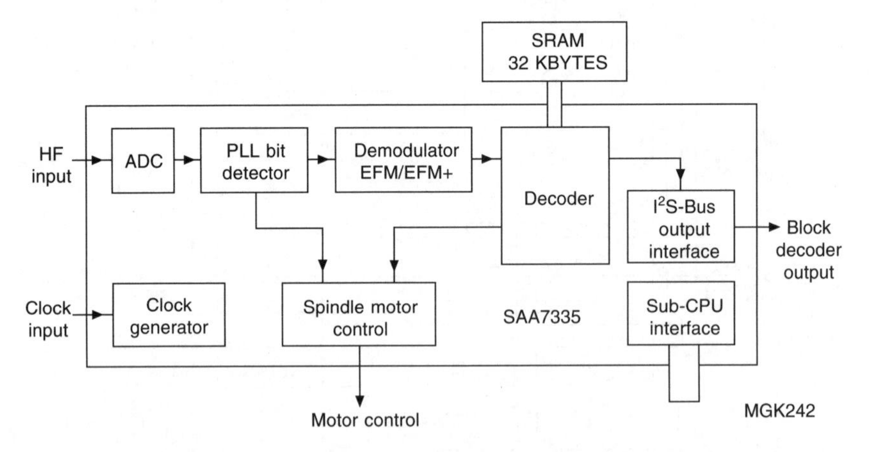

8. Briefly describe the sequence of events that you expect a DVD player to follow when the tray with a disc inside it is closed.

9. Figure E1.2 shows a circuit diagram of a power supply.

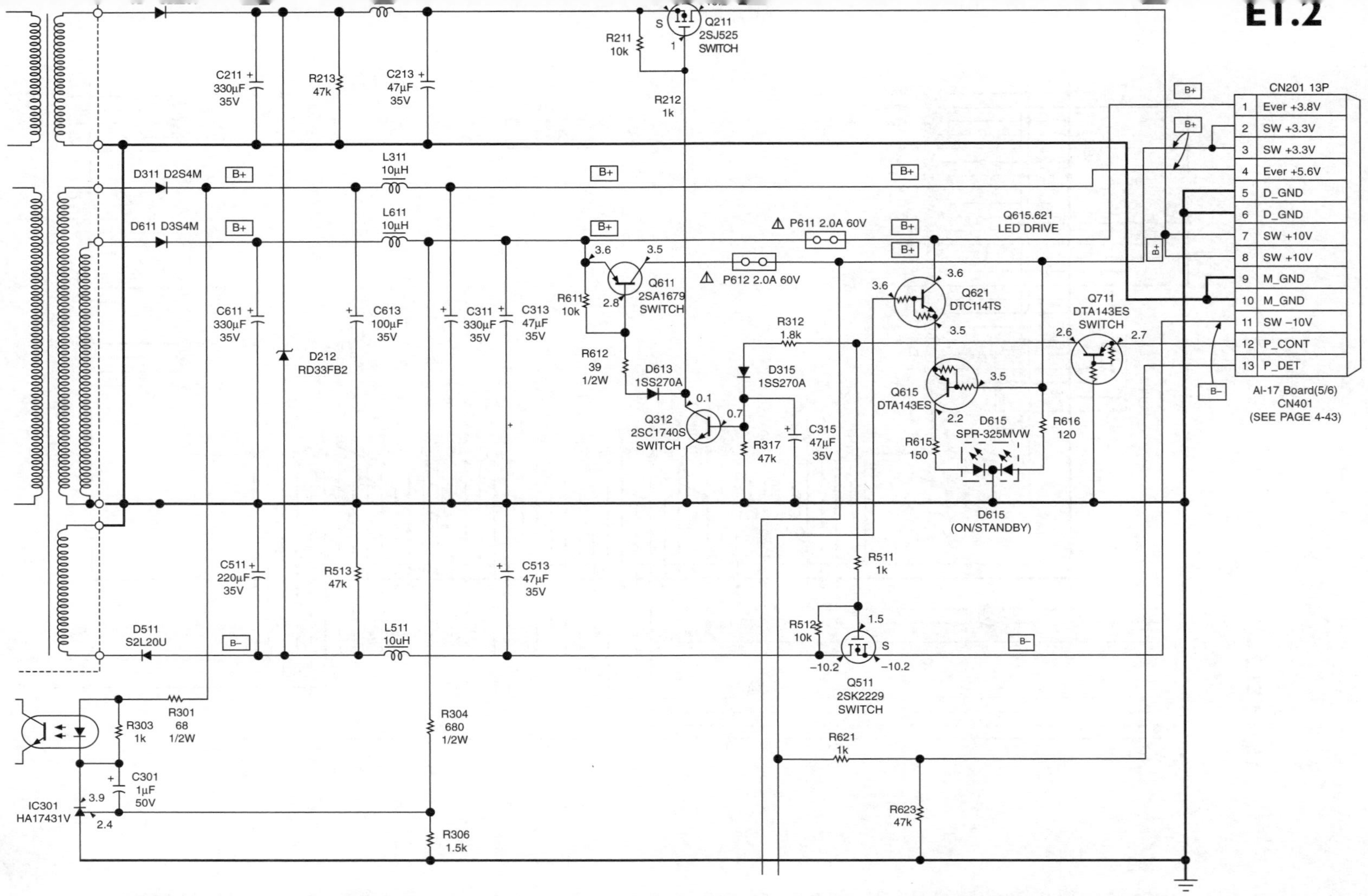

E1.2

CN201 13P
1 Ever +3.8V
2 SW +3.3V
3 SW +3.3V
4 Ever +5.6V
5 D_GND
6 D_GND
7 SW +10V
8 SW +10V
9 M_GND
10 M_GND
11 SW –10V
12 P_CONT
13 P_DET
AI-17 Board(5/6)
CN401
(SEE PAGE 4-43)
Q211
2SJ525
SWITCH
R211
10k
R212
1k
C211
330µF
35V
R213
47k
C213
47µF
35V
D311 D2S4M
D611 D3S4M
L311
10µH
L611
10µH
B+
B–
Q611
2SA1679
SWITCH
R611
10k
R612
39
1/2W
D613
1SS270A
Q312
2SC1740S
SWITCH
P611 2.0A 60V
P612 2.0A 60V
R312
1.8k
D315
1SS270A
R317
47k
C315
47µF
35V
Q615.621
LED DRIVE
Q621
DTC114TS
Q615
DTA143ES
R615
150
D615
SPR-325MVW
R616
120
D615
(ON/STANDBY)
Q711
DTA143ES
SWITCH
C611
330µF
35V
D212
RD33FB2
C613
100µF
35V
C311
330µF
35V
C313
47µF
35V
C511
220µF
35V
R513
47k
C513
47µF
35V
D511
S2L20U
L511
10uH
R511
1k
R512
10k
Q511
2SK2229
SWITCH
R621
1k
R623
47k
R303
1k
R301
68
1/2W
R304
680
1/2W
C301
1µF
50V
IC301
HA17431V
R306
1.5k

Figure E1.3

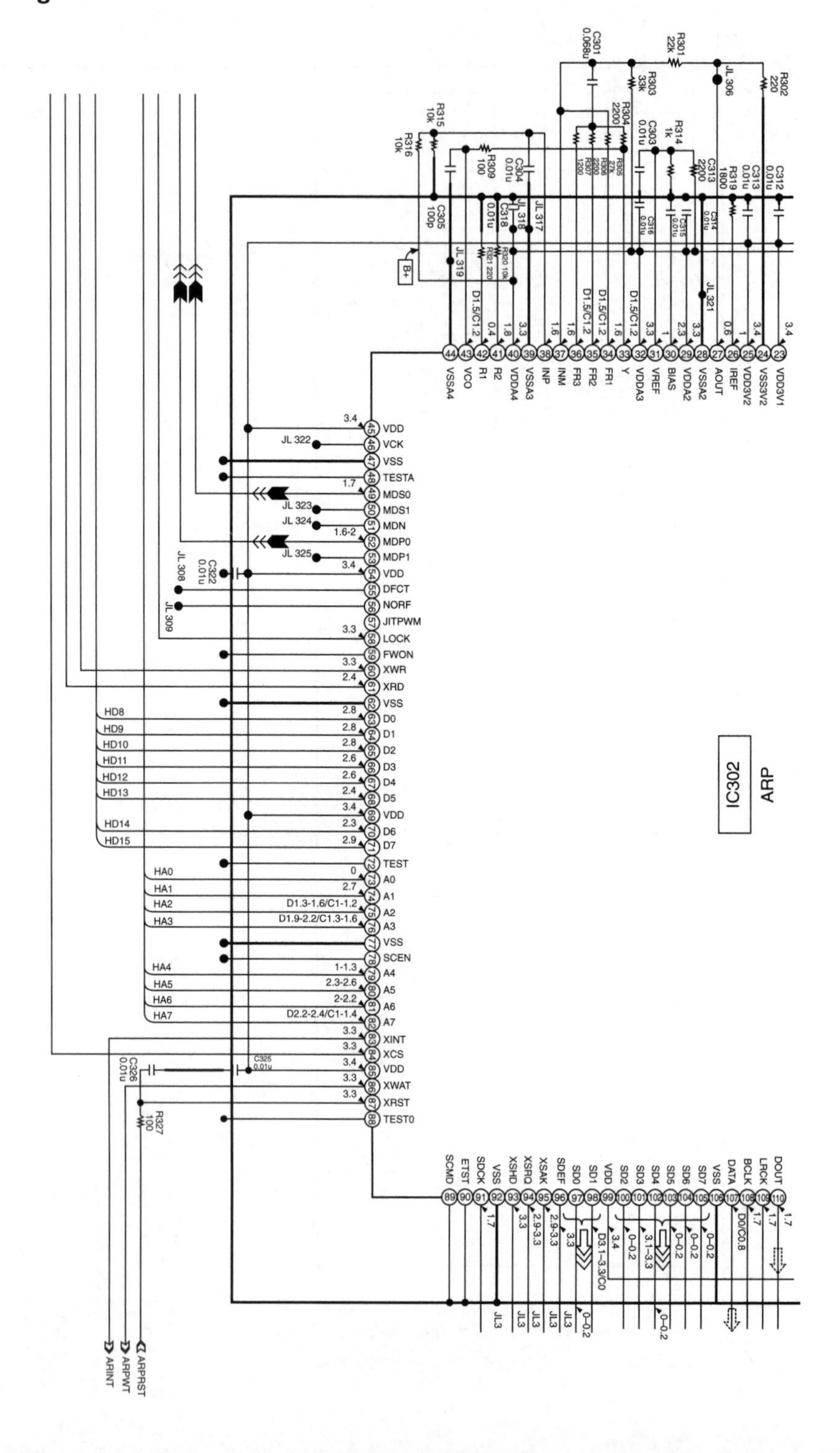

IC302
ARP
VDD
VCK
VSS
TESTA
MDS0
MDS1
MDN
MDP0
MDP1
VDD
DFCT
NORF
JITPWM
LOCK
FWON
XWR
XRD
VSS
D0
D1
D2
D3
D4
D5
VDD
D6
D7
TEST
A0
A1
A2
A3
VSS
SCEN
A4
A5
A6
A7
XINT
XCS
VDD
XWAT
XRST
TEST0
HD8
HD9
HD10
HD11
HD12
HD13
HD14
HD15
HA0
HA1
HA2
HA3
HA4
HA5
HA6
HA7
JL 322
JL 323
JL 324
JL 325
JL 308
JL 309
C322
0.01u
C325
0.01u
C326
0.01u
R327
100
ARPRST
ARPWT
ARINT
B+

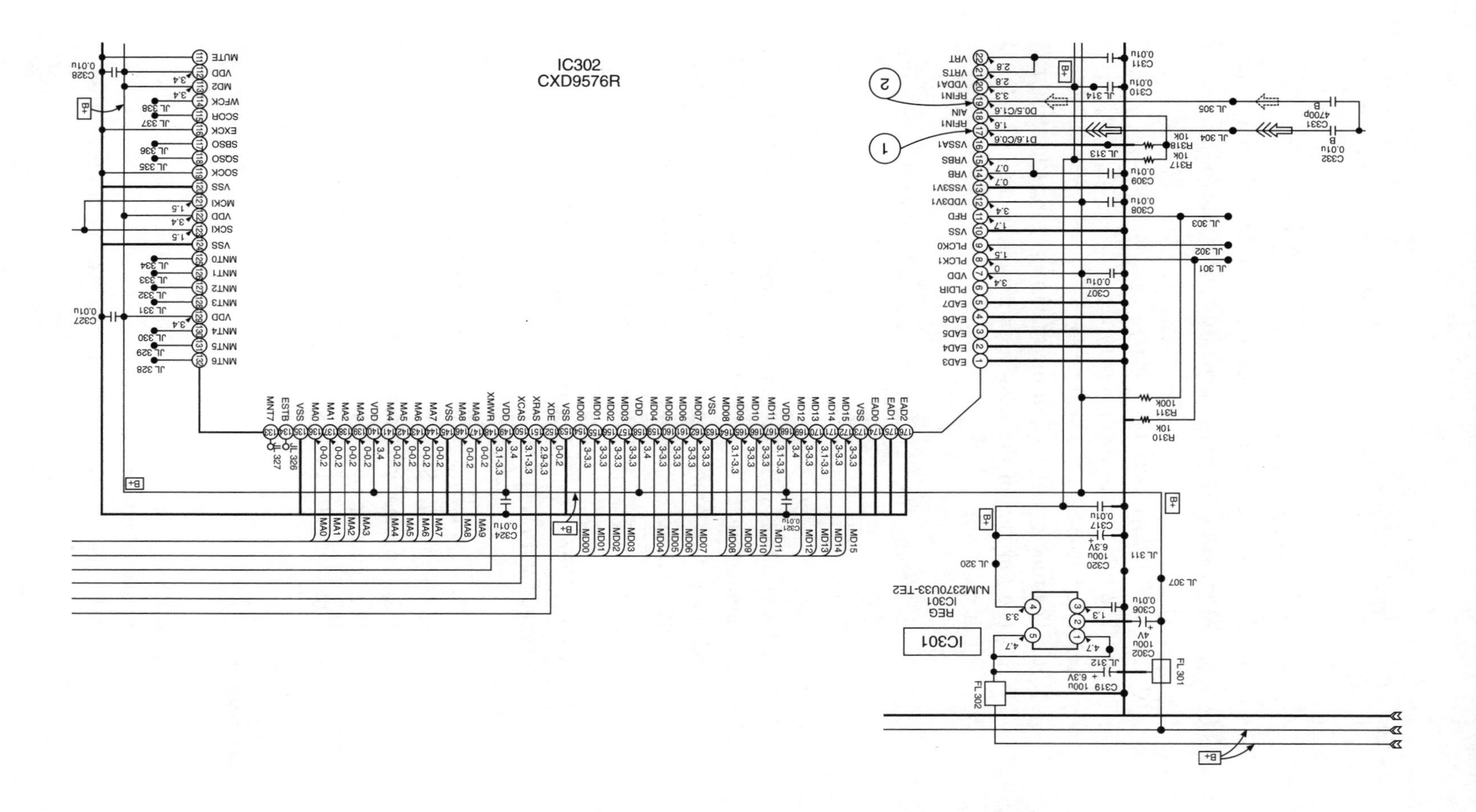
IC302
CXD9576R
IC301
REG
IC301
NJM2370U33-TE2
FL 301
FL 302

a. State the purpose of P_DET and P_CONT
b. State, with explanation, the effect of the following faults: (i) D211 o/c (grid 4A); (ii) R621 o/c (grid 6D).

10. In a DVD player, briefly explain:
    a. The function of the de-encryptor
    b. The purpose of the sled motor
    c. The function of each of the two lines of the I^2C serial bus
    d. The purpose of the lead-in and lead-out area of a DVD disc.

11. With reference to a DVD drive, give a brief explanation of:
    a. IRQ
    b. The driver.

12. A DVD title is to be produced, comprising two elements: a 75-minute music section Dolby Digital (AC3) 5.1, and a 120-minute video section with MPEG-2 stereo audio and a facility for two language subtitles. The title is to be recorded on a DVD-5. Calculate the video bit rate.

13. With reference to DVD writers, state the difference between DVD-R and DVD-RW.

14. Refer to Figure E1.3, which shows an RF processor chip used in a DVD player. State:
    a. The expected waveform at pin 17
    b. The function of pin 147
    c. The function of pin 172.

Answers

1. a. Anamorphic: a 16 : 9 picture that has been squashed to fit an aspect ratio of 4 : 3
   b. AC3: also known as Dolby Digital AC3 is a Dolby surround

sound system with five channels plus one low frequency channel, known as 5.1

c. GOP: Group of Pictures used for the implementation of temporal video compression
d. PCI: Programme Control Information, a PES that contains such information as aspect ratio, language, PAL/NTSC
e. SoC: a highly integrated chip that contains a microprocessor core (normally RISC) and one or more processing applications.

2. a. (i) Temporal compression is an interframe compression that compares successive frames, retaining differences between them; (ii) Spatial compression is an intraframe video compression, which removes unnecessary repetitions of the contents of an individual picture frame
   b. (i) The sampling frequency for luminance is 13.5 MHz; (ii) The sampling frequency for chrominance is 13.5/2 = 6.75 MHz for each component.

3. a. CLV refers to the constant linear speed of the track movement under the pickup head
   b. CBR refers to the rate at which the disc is read by the pickup head, which is constant at 26.16 Mbits/s
   c. Focus search is the process of the pickup head attempting to focus the laser beam on a track, which involves moving the objective lens in and out of focus
   d. The photodetector receives and detects the reflected laser beam off the disc and produces the RF signal, as well as other signals such as focus error, tracking error.

4. a. $2 \times 60 \times 60 \times 3.5/1000 = 3.15$ GB (billions of bytes)
   b. For a double-layer disc, the minimum length of a pit must be slightly longer than that for a single-layer disc.

5. a. (i) Block A is the RF processor; (ii) Block B is the video encoder; (iii) Block C is the video SDRAM

b. Flash is a non-volatile memory store that holds the start-up and other routines, such as video and audio decoding, RF processing, etc.; these routines are called up by the processor as and when required
c. (i) IC805 is the main system processor (sys con), which carries out the initialization, programming and control of all the units in the system; (ii) X801 is a crystal that sets the 20-MHz frequency clock for the sys con chip.

6. a. 20-MHz chip clock pulse
   b. 5-V DC (Reset)
   c. 27-MHz system clock pulse
   d. For PAL, 15.625-KHz line sync pulse
   e. Interrupt request pulse, frequency in MHz.

7. a. The HF input is analogue
   b. The demodulator has two functions: for DVD applications, it converts 16-bit words into 8-bit words; for CD applications, it converts 14-bit words into 8-bit words
   c. The SRAM memory chip stores recording sectors to allow the DSP to de-interleave them into their original order
   d. With a faulty SRAM chip there would be no de-interleaving, resulting in failure to reproduce the original video, audio, etc. PESs – and hence no video, audio, PCI or DSI. The disc will stop rotating and player reverts to standby.

8. When the tray is closed, or when 'playback' is selected, a servo start-up routine is initiated to carry out the following:
   - The spindle motor rotates the disc at a relatively high speed
   - The laser beam is directed onto the disc and the reflected beam is detected by the photodiodes to produce the RF signal
   - The sled motor moves the optical head across the disc from the centre towards the circumference and back again, to determine the diameter size and generate a tracking error signal

- The two-axis actuator moves the objective lens up and down to obtain a focus, a process known as Focus Search.

If the tracking error voltage is low (0.4 V or less), then the inserted disc is a DVD. A CD disc, with its longer pits and wider track pitch, will produce a higher TE voltage of around 2 V.

During the focus search, if the focus error is high (around 1 V), this indicates a highly reflective surface and thus a single-layer disc. Alternatively, an FE of around 0.5 V indicates a dual-layer DVD disc.

9. a. When the power supply is turned on from the mains, P_DET goes HIGH, sending a signal to the user interface that the PS is ready to switch on. The interface sends a signal back in the form of P-CON HIGH to turn the switching transistors Q211 and Q511 on. To turn the player to standby, the user interface brings P_CON down to 0 V, switching Q211 and Q511 off.
   b. (i) The fault D211 o/c means no 10-V supply and hence no motor drives, so the player will always return to standby; (ii) With the fault R621 o/c, there will be no P_DET signal and hence the only voltages supplied by the PS will be the ever voltages (3.8 V and 5.6 V). The PS will remain in the standby mode all the time.

10. a. The de-encryptor removes the encryption or scrambling that was introduced at the encoding stage
    b. The sled motor moves the pickup head to follow the spiralling track
    c. These lines are the data line and the clock line
    d. The lead-in area is the first area of a DVD layer; the lead-out area is the last area of a DVD layer.

11. a. IRQ is interrupt request, a hardware control line for a microprocessor chip, which is activated by other devices when they wish to be serviced by the processor

b. The driver is a software routine required by a processor to control a device like a DVD drive.

12. The starting point is to calculate the available space on the disc that may be used for recording the program:

$$
\begin{aligned}
\text{DVD-5 capacity} &= 4.7 \text{ billion bytes} \\
&= 4.7 \times 10^9 \text{ bytes} \\
&= 4.7 \times 10^9 \times 8 \text{ bits} = 37.6 \times 10^9 \text{ bits} \\
&= 37\,600 \times 10^6 \text{ bits} \\
&= 37\,600 \text{ Mbits}
\end{aligned}
$$

With a 4 per cent disc space reserved for overheads,

$$
\begin{aligned}
\text{Available disc space} &= 37\,600 \times 0.96 \\
&= 36\,096 \text{ Mbits}
\end{aligned}
$$

Starting with the audio stream, the Dolby Digital 5.1 format is usually implemented using a bit rate f 0.384 Mbits/s. The disc space occupied by Dolby Digital 5.1 surround sound can be calculated as:

$$
\begin{aligned}
\text{Audio bit rate} \times \text{length of program} &= 0.448 \times 75 \times 60 \\
&= 2016 \text{ Mbits}
\end{aligned}
$$

Therefore

$$
\begin{aligned}
\text{Free disc space} &= \text{available disc capacity} - \text{space taken up by the audio element} \\
&= 36\,096 - 2016 \\
&= 34\,080 \text{ Mbits}
\end{aligned}
$$

The video section consists of two elements: 120 minutes of MPEG-2 audio, and 120 minutes of video. The video-related audio element requires a bit rate of 0.384 Mbits/s, thus occupying a disc space of $0.384 \times 120 \times 60 = 2765$ Mbits.

The subtitles element requires a bit rate of 0.04 Mbits/s per track or language, giving a total bit rate for the subtitle element of $2 \times 0.04 = 0.08$ Mbits/s.

Consequently, the disc space required for the subtitle element is $0.08 \times 120 \times 60 = 576$ Mbits.

Therefore:

Total video-related elements = 2765 + 576 = 3341 Mbits

Summary of disc usage:

Music section	2016 Mbits
Video-related	3341 Mbits
Total non-video	5357 Mbits

It follows that the disc space available for the video = 36 096 – 5357 = 30 739 Mbits. Thus:

$$\text{Average video bit rate} = \frac{\text{Video disc space}}{\text{Length of video element}} = \frac{30\,739}{120 \times 60} = 4.27 \text{ Mbits/s}$$

13. DVD-R is a recordable DVD, a write-once DVD format. DVD-RW, rewritable DVD, is a format by which a disc may have its contents changed several times.

14. a. The expected waveform at pin 17 is an eye pattern representing the RF signal coming from the optical head if viewed by an analogue oscilloscope. If viewed by a storage oscilloscope, it is an amplitude-varying signal representing logic 0 and logic 1.
    b. Pin 147 is a memory address line for the RF processor memory chip.
    c. Pin 172 is a data memory line for the RF processor memory chip.

Index